Computational Botany

Paolo Remagnino · Simon Mayo
Paul Wilkin · James Cope
Don Kirkup

Computational Botany

Methods for Automated Species Identification

 Springer

Paolo Remagnino
Faculty of Computing, Information Systems
  and Mathematics
Kingston University
Surrey
UK

Simon Mayo
Department of Identification and Naming
Royal Botanic Gardens, Kew
Richmond, Surrey
UK

Paul Wilkin
Department of Natural Capital and Plant
  Health
Royal Botanic Gardens, Kew
Richmond, Surrey
UK

James Cope
Digital Imaging Research Centre
Kingston University
Surrey
UK

Don Kirkup
Department of Biodiversity Informatics
  and Spatial Analysis
Royal Botanic Gardens, Kew
Richmond, Surrey
UK

ISBN 978-3-662-57156-9   ISBN 978-3-662-53745-9   (eBook)
DOI 10.1007/978-3-662-53745-9

Printed on acid-free paper

This Springer imprint is published by Springer Nature
The registered company is Springer-Verlag GmbH Germany
The registered company address is: Heidelberger Platz 3, 14197 Berlin, Germany

Acknowledgements

We thank the Leverhulme Trust (grant F/00242/H) for funding the research project MORPHIDAS: Morphological Herbarium Image Data Analysis, which was carried out by a consortium of researchers from Kingston University, the University of Surrey and the Royal Botanic Gardens Kew, and provided essential support for material presented in this book. We thank Jonathan Clark, David Corney, Sarah Barman, Lilian Tang, Richard White and Ken Bailey for their contributions to the project, and The Board of Trustees of the Royal Botanic Gardens, Kew for access to living and herbarium plant collections.

Contents

Introduction

1

1.1 Plant Species and Their Identification

One of the most immediately obvious features of organic life is its extraordinary diversity — Charles Darwin's "endless forms" (Darwin 1859). Evolutionary biology attempts to understand how nature's diversity arose and the sciences of ecology, genetics and physiology investigate the interactions of organisms and seek to understand how they function as part of higher level systems within the cosmos as a whole. At the bottom of the pyramid, individual organisms are usually self evident (though not always - Wilson 1999), and despite their bewildering variation, a more patient and experienced eye soon discerns that they can be grouped into "kinds", a word which translated into Latin becomes *species*. Species recognition belongs to the domain of biological taxonomy and "recognition" must be understood in two distinct modes; the first is to discern the units and attempt their delimitation — this is called "segmentation" in computerized pattern recognition, and "classification" or "clustering" in taxonomy. The second meaning is to identify or "determine" an organism as belonging to a pre-existing species, i.e. "classification" as understood by computer scientists. Although this book focuses on identification, "segmentation" is a constant companion to our thinking, because the features to be used for identification must be congruent with, if not among, those which delimit the species.

In the broad sense, taxonomy is not unique to biology. In any area of scientific knowledge and indeed perhaps in any area of human intellectual activity, one must begin by defining units of some kind - a process widely known (though not so much in biology) as typology and which results in classifications of the phenomena of interest. Examples are the constituents of matter — the kinds of atoms — first recognized scientifically as the elements in Mendeleev's Periodic Table (Gordin 2004), and languages, which are based on hierarchical units such as phonemes, alphabets and words. Even inherently vague phenomena such as colours and cloud forms have been discretized into colour terms and cloud types. Plant taxonomy has similarly developed from the vernacular typologizing of plants by people over millenia (Atran

© Springer-Verlag GmbH Germany 2017
P. Remagnino et al., *Computational Botany*, DOI 10.1007/978-3-662-53745-9_1

1990), using features which are relatively easy to see and characterize visually. These characters are predominantly those of the external parts of the plant, the sizes, shapes, colours, patterns and textures of its stem, leaves, flowers, fruits and seeds, the totality of which make up the form or "morphology" of the plant (from *morphé*: form, shape, outward appearance, Greek). Other, non-morphological characteristics have also been traditionally used to help characterize species, e.g. habitat, phenology and easily verifiable genetic features, such as the production of progeny which strongly resemble the parents.

From its inception, the science of botany has been founded on the recognition of two kinds of elementary units: the species into which plants are grouped, and the structural components which make up an individual plant. These two areas fundamentally concern the study of patterns and are respectively called plant taxonomy and plant morphology (including anatomy). Each has its cognate, process-oriented research area, respectively plant evolution and plant growth and development, which focus on the transformations of their respective patterns through time.

As a scientific discipline, plant morphology developed more recently than taxonomy, since it required both the amassing of large comparative collections of plant specimens, and technological developments such as microscopy. However, taxonomy and morphology have been historically highly interdependent, in that the definition of a plant taxon depends on the prior definition of the features by which it can be distinguished from other such taxa. As the study of morphology has become more sophisticated, so in turn taxonomists have been able to resolve finer and more complex distinctions in the classification of plant species.

Plant taxonomy itself can be viewed from two rather different perspectives. On the one hand the recognition and investigation of species taxa poses an ontological problem — are these groups "real" and how have they arisen? — the subject matter of evolutionary biology. On the other hand, the existence of a system of plant species is of basic practical importance to humankind because species can be regarded as recognisable groups of individuals with certain properties in common, some of which are of great importance for human life, as characteristic of food, medicinals and products of economic and cultural value.

It is in this context that the importance of the identification of plant species can be appreciated more fully. Identifying a species provides the key to knowledge about its properties (geography, ecology, chemistry, genetics, ethnobotany, etc.). Its identification depends on the prior creation of a system of recognized species (classification), themselves defined on the basis of a system of structural units (morphology: leaves, flowers, stems, roots, etc.).

How do taxonomists define and describe ("segment") species, and then subsequently identify ("classify") individual plants thereof? Basic to this activity is familiarity with the plants themselves and with the catalogue of morphological features used to characterise and distinguish the species. The process of group-recognition — "segmentation"— is largely intuitive and founded in the remarkable pattern recognition capabilities of the human brain. In orthodox taxonomy, defining and recognizing species is a process that requires the intuitive manipulation of many features simultaneously, the result being that a group of individuals can be recognized as having a kind of identity. This identity can be expressed, using a geometrical metaphor, as a high-dimensional space whose axes are the descriptive features observed in the plants. For an individual to be identifiable as a member of a species, the observed group features must each have a characteristic range of values, and those of some features (diagnostics) will be either unique or characteristic of the species to a high degree. This feature subspace is equivalent to the species "taxon concept", using the latter term as described by Berendsohn (2003).

The "segmentation" process is mostly subjective and its details are not accessible to direct observation, although some aspects can be studied, as indicated in the last chapter of this book on botanists' vision. Taxonomists also tend to be specialists because different plant families and life forms have special morphological features which must be learned. Taxonomists' brains are their most important technical equipment and must be trained over a period of time by exposure to the particular data of plants in order achieve optimal performance. And as with most areas of knowledge and activity, some people are just naturally more able than others at defining and recognizing species.

Once the "segmentation" has been achieved and the groups to be formally recognized as species have been set up, identification of new individuals can be carried out. Usually this is done by selecting diagnostic characters from the species descriptions and organizing them in the form of a key (Pankhurst 1991). However, in principle any kind of feature that is congruent with the delimitation of a species can be used as a diagnostic character, and need not come from the set of characters used in the "segmentation" process. Molecular bar-coding (CBOL PlantWorking Group 2009), for example, is a technique for identifying species using a small, diagnostic segment of DNA and uses a set of features which are entirely separate from the (usually) morphological descriptive characters used to delimit the species taxon.

Morphology itself also has the potential to produce new kinds of species diagnostics. Morphometric techniques can deploy enhanced transformations or manipulations of traditional morphological characters for the goal of identification. For example, leaf shape has been used for centuries as a qualitative descriptive and diagnostic character of plant species, but modern geometric morphometrics (see below) has transformed the leaf shape feature into sets of statistically manipulable quantitative parameters, such as vectors of Fourier coefficients or landmark configurations, which can make possible mathematically more powerful comparisons between samples of different species. Many other traditional botanical descriptors are qualitative (multistate, unordered) because this has been the only practical way of using them for taxonomic purposes, e.g. branching patterns of inflorescences, vesture (types of hairiness), leaf venation, external sculpture patterns of seeds and pollen (e.g. Stearn 1992). Each of these characters is potentially quantifiable and capable of yielding new variables that could be used enhance identification through computational methods.

The core topic of this book is the mining of information from plant images in novel ways to discover quantitative parameters which can be deployed as species diagnostics and thereby improve automatic identification systems. The domain of plant morphology is the source of the dataused while plant taxonomy provides the groups (species) to be diagnosed. Since the subject matter of the book is largely methodological, it is unnecessary to explore these two disciplines in any detail. On the other hand, the techniques of analysis of morphological data are a more important focus and this has made it necessary to present a brief review of the research field of morphometrics in Chap. 2.

To be diagnostic, any new measure must (1) be generated from a set of individuals which has been identified as belonging to the species using a pre-existing taxon concept (the training set), and (2) must generate a range of values which characterizes that species to a high degree. But the interplay between the original taxon concept and new diagnostic features is not a simple one. Although the data that new techniques of data capture and analysis might generate from a sample of individuals of a species do not by necessity become part of its taxon concept, there seems no reason why new diagnostics cannot become so, thus progressively enriching it. This problem dogs the discussion and standardization of species limits which, perhaps contrary to the wider public's understanding, are a permanent work-in-progress (Knapp et al. 2005).

Automating identification has become increasingly important during recent decades for many reasons. The most important of these is the constant and unrelenting depletion of the world's biodiversity through human agency. A century ago the most biodiverse regions in the tropics were largely untouched and their taxonomic exploration could be undertaken at a more leisurely pace. Specimens collected on lengthy and hazardous expeditions to remote parts of the world could await the attention of taxonomic specialists for the years and sometimes decades which they needed to prepare their floras, monographs and revisions; the identification of little-known species was viewed as urgent by very few people. The situation is today so completely transformed that the time delay in identifying newly collected specimens or describing newly discovered species has been widely dubbed the "Taxonomic Impediment" by conservation organizations such as the Global Plant Species Initiative (Godfray 2002; Riedel et al. 2013), drawing attention to the shortage of taxonomists worldwide, the main obstacle to providing a more efficient service for the needs of conservation science. The internet has also played a vital part in driving the demand for automated identification, not only for its role as a new medium within which identification can be undertaken, but also by stimulating the development of e-taxonomy (Clark et al. 2009), which is making taxonomists' data far more accessible than ever before. An excellent account of the rationale for automatic identification of organisms is given by MacLeod (2007b) in the introduction to a recent volume dedicated to this subject and dealing mostly with examples from zoology and palaeontology (MacLeod 2007a).

Despite the enormous expansion in the availability of molecular, physiological and genetic data on plant species, our knowledge of plant biodiversity is still based for the most part on herbarium and museum collections the world over. These consist of hundreds of millions of plant specimens accumulated over the past three centuries by constant exploration and survey in the field and preserved in institutions which generally depend on public funding for their continued maintenance and accessibility. These collections have two primary purposes. First, they constitute the material from which the classificatory system of species has been and continues to be created by taxonomists, through species definition and naming. And second, they are used as a reference by means of which newly collected material can be identified to species by matching to specimens which act as standards. This dual role frequently has the result that new species are discovered, leading to expanded knowledge of both biodiversity and past evolutionary processes.

Herbaria and museum collections thus represent the fundamental, distributed database of the world's plant diversity. Many herbaria, such as that of the Royal Botanic Gardens at Kew, London, are now creating internet archives of high quality digital images of their specimens, and this activity seems certain to grow dramatically over the coming decades. We can thus look forward to a time in the near future when there will be a huge array of taxonomic information represented by specimen images, accessible through the internet for various purposes, including automated identification systems.

1.2   The Delimitation of Species by Descriptions

Plant species are based ultimately on written taxonomic descriptions (Winston 1999), which use a large range of specialized terms to represent morphological characters (Stearn 1992). When delimiting species with morphological characters, botanists look for diagnostic characters — ideally those present in all members but absent in non-members. Usually only a minority of characters in a species description are diagnostic. Diagnostic characters which hardly vary within the species are regarded by taxonomists as "good". If two groups overlap in the ranges of all their characters they are normally not regarded as distinct species. The diagnostic characters of species may be non-morphological, such as differences in chromosomes, ecology, phenology, breeding systems, DNA sequences, etc., but in most cases plant species are distinguished by features of their morphology.

The initial stage in delimiting a species is a largely intuitive clustering process which depends on recognizing visual patterns among individuals. The detailed delimitation and description can only be carried out once the boundaries of the species have been established and here familiarity with the plants is a crucial factor.

The species description expresses variation in characters as simple ranges between extremes and is not a statistical summary. Descriptive taxonomy seeks gaps in variation between species, i.e. discrete distinctions. In a statistical approach the focus would be on describing the probability distributions of the characters and their central points, e.g. their means. Even this were the aim, it would be confounded by the available specimens — the source of most data — since these are haphazard assemblages and hardly ever result from planned random sampling; besides this, only few specimens may exist for a given species.

This descriptive procedure is highly synthetic. The original data points - the vectors of observations obtained from each individual plant - cannot be recovered from the final published description, nor can the variation in characters (discrete or continuous) be quantified effectively using the description as a starting point. Nevertheless, taxonomic species descriptions continue to provide the foundational units for knowledge of biodiversity. The descriptions of the approximately 400,000 currently accepted plant species each delimit a taxon concept (Berendsohn 2003), which can be understood as a feature space, and this vast edifice of units is our basic map of plant diversity.

The diagnostic characters of plant species often include features of the leaf blade but equally often features from other plant organs (e.g. life form, stem, inflorescence architecture, flowers, fruits, seeds), and observation of such characters may require examination by low power microscopy. It is also not unusual that the diagnosis of a species includes no characters of the leaf blade. A "gold standard" automated identification system would deploy all the diagnostic characters used by the taxonomist and would test a new specimen against the feature space of the full range of diagnostic characters of candidate species. Only then could a test specimen be said to have been classified using the characters determined as diagnostic by the taxonomist.

1.3 Using Leaves to Diagnose Species

Major obstacles stand in the way of achieving a "gold standard" automated identification system. Data is needed from a potentially very wide range of organ features which differs from species to species, and often would require human intervention to obtain — e.g. dissection and recognition of small parts. An automated system would need to be able to distinguish the different plant organs and the states of each corresponding to the descriptive terms used by botanists, since these will often be diagnostic. The system would also have to be capable of measuring the quantitative variables used by taxonomists, e.g. count leaf lobes or stamens, measure petal widths and so on.

Automated systems have instead approached the problem by focussing on leaf blades, since they are more widely available, easy to image, and can mostly be regarded as two-dimensional without serious distortion. The assumption is that most species can be distinguished by leaf features and this is reasonable since leaf blades are morphologically very diverse and very frequently used as the source of diagnostic characters (Fig. 1.1). However, as previously mentioned, leaf features may not in fact be diagnostic in a given species according to its taxonomic delimitation and so there is no *a priori* guarantee that a given species can be identified using leaf features alone.

This is not a new problem. Field identification of tropical forest trees is made extremely difficult in the absence of flowers or fruits, which are often unobtainable or unavailable. Identification and description of fossil plants must often be done using leaves alone since it is unusual to find fossil leaves and flowers or fruits incontrovertibly interpretable as belonging to the same plant and leaves are far more frequent as fossils. In both these research areas botanists have studied leaf characters in great detail in order to increase their diagnostic potential (Hawthorne and Jongkind 2006;

Fig. 1.1 The main components of a typical leaf

Hickey 1973; Ellis et al. 2009) and as a result have produced a wealth of new information for species identification.

Just as taxonomists have focussed on details which are most tractable in the physical examination of herbarium and living specimens, so the automatic processing of information from leaf images focusses on those features which are most easily captured by automated technology. The two are often very different. It is easy for computer processing to make quantitative measurements of characters which are very troublesome by hand, such as leaf area, leaf perimeter (neither used in orthodox descriptions) and as will be seen in later chapters, automatic data capture from leaves involves the creation of a range of entirely new features that have never previously been used by botanists.

What underlies the automatic approach is the existence of "training sets", i.e. sets of leaves previously allocated by taxonomists to the correct species using botany's traditional classificatory "language" of recognized organs, and a large set of descriptive terms describing the states observed among plant species. Given a taxonomic partition of the study specimens into a set of equivalence classes (the species), automated systems are then able to harvest a potentially limitless range of feature data and study their utility as diagnostic characters. Automatic processing also holds out the prospect of gathering data sets which are orders of magnitude larger than those normally used by taxonomists, thus potentially revealing variation patterns that have gone undetected hitherto.

This highlights the importance of the taxonomists' partition — the set of "bins" which must exist prior to the computation of diagnosability and "within-bin" variation in the features under study. However, the bins themselves are subject to change as knowledge of the world's flora improves. New observations from previously unstudied regions can change the boundaries of species and so in the future, automated identification systems will need to cope with changes made by taxonomists in the configuration of the training sets.

Computational plant taxonomy as presented here is in a foundational phase and its main data source is the plant leaf. However technological advances will surely allow other plant organs to act as data sources in the future and eventually systems will be possible that carry out identification on a broader base of biological information than just leaf blades.

1.4 Computational Botany

The automatic identification of plant species, the focus of this book, can be seen to be one aspect of computational botany. There are, however, other research areas which can lay claim to this title and they will be briefly mentioned here.

Using computational methods for helping to define species taxa has been underway since computers became widely available to research scientists from the 1950s onwards. This was when "numerical taxonomy" (also termed phenetics) first flourished and soon developed a range of computational methods for analysing multivariate taxonomic data (Sneath and Sokal 1973; Sneath 1995). One major objective of the numerical taxonomists was to establish a more objective basis for the recognition of taxa, including species. This computational research area can be thought of as attempting the automated definition ("segmentation") of taxa, and it includes the subsequent and related development of morphometrics and cladistics, which are discussed in more detail in the next chapter.

A third area of computational botany is that of "virtual plants"; more precisely, the modelling of plant structure, growth and development by dynamic computer graphics. This can be regarded as the computational analogue of plant development and indeed of plant morphology itself. Here we give only a very brief note on this large research field with a few pointers to the literature.

Modelling 3-D plant structures has been addressed by various research groups working in different areas of plant science such as forest ecology, remote sensing, agricultural science and horticulture and is becoming increasingly important as a technological tool for prediction of change in plant-covered landscapes over time.

The initial development of computer models to simulate the modular growth of plants (e.g. Honda 1971; Frijters and Lindenmayer 1974) began at around the same time as the conceptualization of models of tree architecture by Hallé and Oldeman (1970), later augmented by Hallé et al. (1978). Related advances in plant population biology and morphology (e.g. Harper 1977; White 1979; Bell and Tomlinson 1980) further helped to lay the foundations of virtual plant modelling, which soon moved on from simulating the growth and development of individual plants to that of plant stands, communities and eventually entire landscapes.

One prominent group of virtual plant researchers, based at the University of Calgary, Canada, features the work of Prof. Przemyslaw Prusinkiewicz and colleagues (e.g. Prusinkiewicz and Hanan 1989; Prusinkiewicz and Lindenmayer 1990; Room et al. 1996; Dale et al. 2014) whose research arose from the theory of L-system fractals created by A. Lindenmayer. They have developed the Virtual Laboratory and L-Studio software systems for plant modelling (http://algorithmicbotany.org/).

Other approaches to plant computer-modelling are represented by research groups centred on the University of Montpellier, France, deriving their inspiration from the innovative work of Prof. Francis Hallé, who focussed originally on understanding the structural diversity of tropical forests (Barthélemy and Caraglio 2007). One of the products of their work is AMAPstudio, an online software studio for modelling plant architecture (Barczi et al. 2007, http://amapstudio.cirad.fr/).

These examples give some idea of the extent and sophistication of modern virtual plant modelling. Detailed recent reviews with more extensive literature citations are given by Prusinkiewicz and Runions (2012) and by Sievanen et al. (2014) and associated papers.

1.5 Aims and Objectives

This book aims to contribute to the young discipline of computational botany by exploring techniques for feature extraction, analysis and classification of leaves, leading to novel techniques and a framework for a robust, automated plant identification system.

There are several steps needed to achieve these goals. The first is to extract and describe the required information from leaf images, such as overall shape, margin characteristics, texture, venation patterns, and venture, i.e. presence and type of hairs. These features must extracted accurately from a leaf and then used to create adequate descriptors for comparison of newly sampled leaves with those from known species.

A second important challenge for automated species identification is presented by the inherent within-species variation in leaf morphology, which in some morphological features can be greater than between-species variation, resulting in significant overlap. The requirement here is to develop machine learning techniques capable of taking this variation into account.

A third step is to combine the different modalities (i.e. leaf components and feature-sets). While it may be possible to classify leaves with reasonable accuracy using single components, it seems obvious that there is great benefit in using multiple components and for this an appropriate framework is required. Furthermore, due to inter- and intra-species variation, certain components may be appropriate for some species but not others. A subsidiary objective is therefore to allow automated selection of components and feature-sets on a leaf-by-leaf basis.

A final objective is to learn how professional botanists perceive leaf images and to apply this knowledge to the problem at hand. This involves studying differences in performance between botanists and non-botanists with eye-tracking technology, and investigating how this information could be applied in an automated system.

Morphometrics: A Brief Review

In the Introduction (Chap. 1) an account was given of the rationale for using the morphology of leaves as the source of raw material for automatic systems for the identification of plant species. In this chapter we focus on how morphological features have been used in statistical analysis to define taxa like species, test pre-existing species classifications and carry out identification. Our aim here is to survey briefly the progress and development of the statistical analysis of morphological variation. This general area of scientific research is called "morphometrics", a term which is a compound of the classical Greek words *morphé*: "form", and *metron*: "that by which anything is measured" (http://www.lexilogos.com/english/greek_ancient_dictionary.htm). "Morphometrics" can thus be construed as "the measurement of form". "Form", in turn, can be understood in a narrow sense as referring only to shape, or more broadly as including shape, structure (which includes size, architecture and internal anatomy) and other aspects of external appearance. It is in the broad sense that we will use it here.

There are a variety of "schools" of morphometrics which are "characterized by what aspects of biological 'form' they are concerned with, what they choose to measure, and what kinds of biostatistical questions they ask of the measurements once they are made." (Slice 2005). The term "morphometrics" seems to have entered the biological vocabulary in an eponymous article by Blackith (1965), but has been used widely only from the 1970s onwards, following the publication of Blackith and Reyment's book "Multivariate Morphometrics" (Blackith and Reyment 1971). Under this title these authors emphasized the application of multivariate statistics to data consisting mainly of linear measurements and hence constituting relative and absolute measures of size, from which shapes of morphological structures could be estimated indirectly.

In the 1980's and 1990's the field of morphometrics underwent a major upheaval with the development of geometric morphometrics (Rohlf and Marcus 1993; Adams et al. 2004), which focusses on the analysis of shape variation in organisms and

© Springer-Verlag GmbH Germany 2017
P. Remagnino et al., *Computational Botany*, DOI 10.1007/978-3-662-53745-9_2

their component parts. The impact of geometric morphometrics has been such that the term "morphometrics" is today widely understood to mean the study of shape variation alone (Zelditch et al. 2012).

The increasing use of automated methods of data capture from the morphology of organisms, either directly or from images, brings into focus other aspects of form which do not seem satisfactorily understood as either shape or size features, e.g. textures (Lexer et al. 2009), patterns visible on the surface such as venation (Plotze et al. 2005) and others mentioned in the following chapters. The analysis of such features and the use of their patterns for identification can also be treated under a concept of morphometrics that is broad enough to include the algorithmic analysis of any aspect of the morphology of organisms (Gaston and O'Neill 2004; O'Neill 2007).

2.1 Historical Background to Morphometrics

In this section the development of morphometrics is briefly outlined, from the biometricians of the nineteenth century to the use of multivariate statistics for the quantitative analysis of morphological features of organisms, an area of study usually known as "Multivariate Morphometrics" (Marhold 2011) or "Traditional Morphometrics" (Marcus 1990), and finally to the rise of geometric morphometrics.

2.1.1 Phase 1: Classical Taxonomy, Evolutionary Theory and Biometry: The Background for Morphometrics

The recognition of species of plants and animals by their appearance (morphology) and behaviour goes back to humanity's own origins and certainly even before that (Atran 1990). Many religions account for the diversity we perceive by means of creation myths and implicit in these is the notion of stasis – that once created, species remain the same, since offspring almost always resemble their parents closely. Contrary to what has sometimes been argued (e.g. Hull 1965a, b), the dependence of the morphological similarity of the individuals of a species on heredity has always been recognized by taxonomists since at least Aristotle (Richards 2010; Wilkins 2009, 2010) and from time immemorial by ordinary people. Nothing is more obvious than that like breeds like.

Prior to the widespread acceptance of evolutionary theory following Darwin's *Origin of Species* (Darwin 1859), taxonomists attempted to delimit distinct units of biodiversity (as species) using qualitative morphological characters, inferring at the same time that the individuals composing these units would represent parent-offspring lineages (Hennig 1966, tokogenesis; Wilkins 2009), thus accounting for their phenotypic similarity. Individual variation, according to this generally accepted view (de Candolle 1813), was just an observed but trivial fact, again intuitively obvious to us all from our own family experience. "Good" species could only be accepted as such if delimited qualitatively; for those organisms that proved more intractable taxonomically, there was an assumption that the necessary characters for achieving satisfactory groups must exist, but are simply hidden from current scientific view. This search still motivates most taxonomists and with good reason, since such species delimitations better serve the practical purpose of dissecting the bewildering diversity of nature into units which can serve as raw material for biological science. Here lies a fundamental tension of taxonomy with the evolutionary worldview: while evolution can be shown to produce groups of phenotypically distinct individuals (species), there is no logical justification for expecting qualitative distinctions between them to necessarily exist, given the theory of evolution – qualitative taxonomic distinctions are sufficient, but not necessary, for the evolutionary delimitation of species. On the other hand, all sciences need units and in biology the fundamental units above the level of individuals are species. Mankind hates the continuum and it has been and remains the taxonomists' lot to discretize biological diversity. In order to do this, taxonomists have always used morphological information and in the process have had to find ways of dealing with the observed fact that morphology varies from one individual to another.

Darwin's theory of natural selection changed the explanatory context of biodiversity in fundamental ways. In particular, he focussed on the significance of phenotypic variation in species and argued that it was variation itself which provided the raw material for the evolution of contemporary species from ancestral ones. An immediate consequence of this was that gradual phenotypic change must occur as a species is transformed by selection into another and that therefore there could be no fundamental distinction between taxonomic varieties and taxonomic species. Continuous variation in morphological characters now achieved new prominence since this would be an expected feature of the evolution of species.

The measurement and the quantitative relations of patterns of phenotypic (mostly morphological) variation and heredity became objects of interest and the detailed study of quantitative variation within and between species and infraspecific groups gave rise to the science of biometrics (statistical analysis of biological data), among whose pioneers were Francis Galton (e.g. Galton 1889), W.F. Raphael Weldon (e.g. Weldon 1889) and Karl Pearson (e.g. Pearson 1901; Pearson et al. 1901). Galton invented the correlation coefficient to help him investigate the principles of inheritance (mainly in people) in the pre-genetics era. These scientists discovered in biological organisms the common occurrence of such patterns as the Normal (or Gaussian) frequency distribution in many variables, both continuous and discontinuous. In order to analyse the data they collected, they made fundamental advances in mathematical statistics, developing a range of new mathematical methods, such as regression, which soon had wide application both within and beyond biology (Briggs and Walters 1997; Porter 1986).

2.1.2 Phase 2: Genetics and Statistical Methods in Evolution, Agronomy and Biosystematics

With the birth of genetics in 1900 following the rediscovery of Gregor Mendel's work, a more complete and satisfactory theory of heredity became available. Mendel (Mendel 1866) discovered his laws of inheritance using a statistical treatment of seven categorical (qualitative) morphological characters of the pea plant (*Pisum sativum*). But major difficulties remained in applying the particulate Mendelian theory to explain the continuous character patterns studied by the biometricians. Important insights were provided by the quantitative work of e.g. Johannsen (1909), Nilsson-Ehle (1909) and East (1915), which clearly distinguished genotype patterns from those of the expressed phenotype (Mather and Jinks 1971). It was Ronald A. Fisher's seminal paper of 1918 (Fisher 1918) that showed that continuous variation of characters was compatible with Mendelian particulate inheritance and he introduced the key concepts of variance and its analysis by partition — Analysis of Variance or ANOVA.

The development of biological statistics underwent rapid and continuous growth in succeeding decades, driven in particular by the application of genetic theory to agronomy and evolutionary theory, resulting in the sciences of quantitative genetics (e.g. Falconer 1989) and population genetics (e.g. Fisher 1999; Hamilton 2009), using data which in the case of plant studies was derived from phenotypic and mostly morphological features of plants. The three mathematical biologists who contributed most significantly to this development, especially between the 1920s and 1960s, were John B.S. Haldane, Ronald A. Fisher and Sewall G. Wright, whose work greatly expanded the scope and power of statistics as a toolkit for analysis of biological variation, both genetic and phenotypic. Reyment (1996) gives an illuminating account of some of the key biometric personalities in the period from Galton to Fisher.

The first half of the twentieth century was an interesting time in the prehistory of morphometrics. It was when genetics was still young and taxonomy still fairly dominant in botanists' thinking. Geneticists' discoveries of the depth and complexity of genetically influenced phenotypic variation within taxonomists' species had a major impact – an enormous amount of quantitative morphological work was done in this period. Turrill's discussion of the ecotype concept (Turrill 1946) gives a flavour of how the influence of genetics on taxonomy was seen at the time, from a botanist steeped in taxonomy but highly sympathetic to the "new systematics" that arose in response to genetic and evolutionary studies (Huxley 1971). This was when the science of biosystematics was born — essentially the quantitative study of infraspecific diversity, leading to the formulation of theories of microevolutionary processes — dubbed the Modern Synthesis (Briggs and Walters 1997).

Stebbins (Stebbins 1950, pages 13–21) gives an excellent survey of the use of quantitative methods in descriptive systematics in the period 1920–1950 (though mostly restricted to Anglophone authors). These studies include Woodson Jr. (1947) on leaf morphometrics of *Asclepias tuberosa*, Gregor et al. (1936) on experimental gardens for biosystematics, Lewis (1947) on leaf variation in *Delphinium variegatum*, Erickson (1943, 1945) on *Clematis fremontii* var. *riehlii*, Fassett (1941) on

Rubus; Fassett (1942) on *Diervilla*, McClintock and C. (1946) on *Teucrium*, Anderson and Whitaker (1934) on *Uvularia*, Anderson (1936a) on *Iris*, Clausen, Keck and Hiesey (Clausen et al. 1940, 1945) on *Potentilla*, Fassett (1943) on *Juniperus*, Epling (1944) on *Lepechinia*, Anderson (1946, 1949) on introgression, Fisher (1936) on *Iris*, Anderson and Abbe (1934) on Betulaceae, Davidson (1947) on polygonal graphs for simultaneous presentation of several variables, Clausen (1922) on *Viola* and Anderson (1936b) on the hybrid index.

Edgar Anderson (Stebbins 1978; Heiser 1995) was a plant geneticist whose work well reflects the impact that genetic understanding had on classical plant taxonomy, e.g. in his studies of introgression (Anderson 1949). He advocated more detailed quantitative morphological studies of plant populations (Anderson and Turrill 1935; Anderson 1941) on the basis that classical taxonomy provided only a very sketchy idea of the true nature of morphological variation within species. He was also innovative in inventing techniques for tracking the occurrence in natural populations of several genetically determined morphological characters simultaneously (Anderson 1949), e.g. his much-copied pictorialized scatter diagrams. This was a way of graphically labelling the points in a scatter plot so that the patterns of more than two characters could be seen simultaneously. These were purely graphical ways of handling multivariate information. His hybrid index was another non-statistical but multivariate technique for handling a larger number of characters with the aim of making rough estimates of the frequency of hybrids in natural populations by a rapid inspection of several characters simultaneously.

2.1.3 Phase 3: Multivariate Statistics and Morphometrics

The problem of clustering individuals (classification *sensu* Gordon 1999) using several to many variables in a more formal (algorithmic) way did not attract wide attention from mathematically-minded biologists until the 1950s, with the birth of numerical taxonomy and the invention of electronic computers. This phase led directly to the development of morphometrics as we currently understand it.

Until then, statistical methods had been used to investigate quantitatively the patterns of discrete and continuous characters mainly in connection with genetic studies. In these the focus was on the hereditary transmission of features, whether for the purpose of breeding and improvement of domesticated plants and animals, or for the purpose of understanding evolutionary processes at the fine scale. The taxonomic context of species as the key units was for the most part taken as given, although the new biosystematic studies often provided evidence of complexity which could confound or at least blur the species delimitations of traditional taxonomic revisions. The important point here is that the focus was on the genetic transmission of individual traits rather than attempting the discrimination of groups of individuals described by several or many characters simultaneously – i.e. multivariate analysis aimed at taxonomic delimitation.

Nevertheless, progress in genetics and the realization that a continuous phenotypic trait was very often influenced by more than one gene, led to methods like multiple regression in which the variation of a single dependent variable is analysed in relation to several or many independent ones. In the 1930s other important methodological foundations were laid for multivariate analysis. Fisher (1936) invented the discriminant function using four different continuous measured flower characters from E. Anderson's genetical work on North American irises (Anderson 1936a), now one of the world's most widely used test data sets in statistics. Hotelling (1931) provided a multivariate test of the difference between groups by generalizing Student's T as Hotelling's T^2 and developed canonical correlation analysis for comparing two data sets (Hotelling 1935, 1936). Mahalanobis (e.g. Mahalanobis 1936) developed the D^2 generalized distance metric, an important measure of multivariate distance which takes into account covariation between the variables (Dasgupta 1995). For Reyment et al. (1984), Mahalanobis's distance "provides the only realistic measure of multivariate distance."

Although originally presented mathematically by Pearson et al. (1901), principal component analysis (PCA) also began to be developed at this time (Reyment et al. 1984), impelled by a focus on the study of allometry, on which Huxley (1932) had published a pioneering monograph. Multivariate algebra using linear dimensions as variables dominated efforts to analyse shape quantitatively. Teissier (1938) was a key early explorer in using PCA for a multivariate approach to allometric changes and this was later developed further by Quenouille (1952), Jolicoeur and Mosimann (1960) and Pearce (1965), among others (see discussions in Sprent 1972; Reyment et al. 1984, 1985, 1991, 1996).

A key publication in the emergence of morphometrics as a distinct discipline was the book *Multivariate Morphometrics* (Blackith and Reyment 1971; second edition Reyment et al. 1984). During the period covered by the two editions, numerical taxonomy was an important area of development and debate in taxonomy and to some extent this is evident in these texts, where morphometrics was seen as just one element of radical changes taking place in the methodology of taxonomy at that time. In the 1960s the numerical taxonomy movement (e.g. Sneath 1961; Sokal and Sneath 1963) had an almost revolutionary effect on taxonomists and during this period, as in the 1970s and 1980s when the cladistics movement became ascendant, almost every facet of taxonomic theory and practice came under scrutiny, e.g. Cain and Harrison (1960), Sneath (1961), Sokal (1961), Cain (1962), Sneath (1962), Sokal (1962), Camin and Sokal (1965), Mayr (1965), Farris (1967), Rohlf (1967), Gabriel and Sokal (1969).

Morphometrics can thus be viewed, at least in its earlier phase, as a subset of a much broader enterprise to quantify the procedures of taxonomy. Both numerical taxonomy and cladistics aimed at a general theory and praxis for taxonomy to replace traditional procedures and to take advantage of the calculating power of (electronic) computers which had then only recently started to become available to researchers. They were concerned with procedures to generate taxonomies at all levels and most discussion and controversy at this period centred on the use of discrete morphological character information for this purpose.

Morphometrics — a term introduced by Blackith (1965) and further popularized by Blackith and Reyment (1971) and Pimentel (1979) — has always been concerned with applying multivariate statistical methodology to the analysis of morphological data sets, using such techniques as principal component analysis, factor analysis, principal coordinate analysis, discriminant function analysis, canonical variate analysis, canonical correlation analysis, non-metric multidimensional scaling, correspondence analysis, etc. Until the advent of geometric morphometrics, the morphological data usually consisted of linear and sometimes angular measurements, i.e. continuous (real number) variables. Numerical taxonomy and morphometrics both arose out of a need to quantify differences between groups of organisms described with multivariate data, i.e. vectors of values of many characters for each organism. Measures of resemblance were developed such as association coefficients and distance coefficients, which involved the transformation of a set of original data, be they continuous or discrete or even non-numeric, into real-numbered scalar values representing the degree of resemblance between every possible pair of organisms in the study. Gower (1971) (see also Gower 2008) developed methods for combining discrete and continuous variables to produce distance matrices which could be analysed by scaling methods, especially principal coordinate analysis (Gower 1966) and this provided a bridge between studies using discrete data and those dependent on continuous data.

Subsequent developments, including the rise of cladistics, phylogenetic taxonomy and molecular systematics, overtook numerical taxonomy as a methodology for constructing the taxonomic hierarchy. Multivariate morphometrics (often termed "traditional" morphometrics, Marcus 1990) has always been more relevant for studies of the lower taxonomic categories and is most used by taxonomists for exploring complex patterns of variation at and below species level. However, taxonomy is by no means the only, nor even the most important area of application of morphometrics. Reyment et al. (1984) discuss a wide range of examples of the application of multivariate morphometrics to biological problems and the computer package PAST and associated literature (Hammer et al. 2001; Hammer and Harper 2008; Hammer 2012) illustrate a diversity of practical applications. A major use of morphometrics has always been the correlation of morphometric variation patterns with those provided by other, non-morphological data, e.g. in ecological and palaeontological studies, such as soil characteristics, chemical composition of rocks, time and space, etc.

2.1.4 Phase 4: Geometric Morphometrics

"Whether broadly ... or narrowly ... construed, morphometrics clearly has something to do with the assignment of quantities to biologic shapes" — thus begins an important review of the field in the early days of geometric morphometrics (Bookstein 1982), a major transformation of morphometrics that took place in the late 1970s and 1980s which resulted in a new field of research (Mitteroecker and Gunz 2009). A recent text is Zelditch et al. (2012) and MacLeod's (MacLeod 2012a, b) beautiful and explicative PalaeoMath webpage is an essential source. The most important feature of this "revolution" (Rohlf and Marcus 1993) was to use configurations of landmark coordinates as the basic data, rather than linear measurements. Although many traditionally used measurements were made between points on the organism that could be considered biologically homologous by the accepted criteria of homology (e.g. Rieppel 1988), once gathered this data no longer preserved the geometric relationships of those points nor their covariation. Landmarks became the new name for selected homologous points on the organisms or biological structures under study and the data consisted of the configurations of the set of landmark coordinates recorded for each study object, each configuration consisting of the same number of landmarks.

A crucial advance that paved the way for geometric morphometrics was a clearer definition of shape: "Shape is all the geometrical information that remains when location, scale and rotational effects are filtered out from an object" (Kendall 1977; Dryden and Mardia 1998; Kendall et al. 1999). This resolved the problem of differentiating size and shape parameters that had plagued traditional morphometrics and with the general acceptance of centroid size as the most suitable measure of size, the way was open to define geometric morphometrics as the quantitative study of biological shape and its variation.

As recounted by Bookstein (1993), the solution to the problem of how to place the statistical analysis of landmark coordinates on a firm mathematical foundation was the result of independent theoretical work by C.R. Goodall, F.L. Bookstein and D.G. Kendall, synthesized in Bookstein (1986) and Bookstein (1991). Fundamental to this was the development of appropriate ways of "filtering" location, size and rotation from the landmark coordinate data so as to produce data matrices containing information only on shape variation. The solution was general Procrustes analysis (see Zelditch et al. 2012 for a recent account). Then Bookstein (1989), in an inspired move, introduced the thin-plate spline (TPS) metaphor from physics to obtain a mathematical interpolation for the space between the landmarks. Thus, in a stroke, the problem of operationalizing D'Arcy Wentworth Thompson's famous morphological transformations was solved (Thompson 1917, 1942). Bookstein derived parameters — principal warps, partial warps and later relative warps — which provided the tools for statistical analysis of the "bending energy" that measures the deformations represented by the study objects when these were considered as shape deviations from the mean shape of the study data set (Bookstein 1991). One reason why the TPS approach has been so influential is because it communicates visually the results of morphometric analysis by reconstructing shapes derived from analysis, e.g. mean shapes of populations, or trends in shape along principal component axes — this was something that multivariate morphometrics had never achieved.

The solid theoretical foundation of geometric morphometrics using landmarks has made it the preferred approach for morphometric analysis in studies where the organisms and their component parts offer sets of biologically meaningful landmarks. This has made the extension of such studies into three dimensions relatively easy especially in anthropological research (e.g. Weber and Bookstein 2010). Bookstein (1991, page 63) gives a detailed discussion of landmark types.

Landmark geometric morphometrics is limited by the need for biologically meaningful homologous points on the structures of interest, but these are often few or lacking in leaves, still the focus of most botanical studies. An alternative approach that developed alongside landmark studies was the analysis of outline shapes using various kinds of functions, including Fourier's harmonic series. Although less well-founded theoretically, outline analysis has nevertheless been important in the absence of alternatives (Lestrel 1997). Recently, however, semilandmarks have come into prominence as a means of capturing the shapes of homologous structures which exhibit digitizable contours between homologous landmarks (Silva et al. 2012; Zelditch et al. 2012; MacLeod 2013; Gunz and Mitteroecker 2013). They have the advantage of making it possible to use the landmark approach, with all its mathematical advantages, to capture the shapes of contours. Zelditch et al. (2012) note that semilandmarks have less degrees of freedom than landmarks and this needs to be borne in mind in analysis. Rohlf's tpsRelw software for analysing partial and relative warps (Rohlf 2010b) performs a sliding procedure which optimizes the position of semilandmarks in relation to the bending energy of the deformations (Bookstein 1997).

Elliptic Fourier analysis (EFA) and Eigenshape analysis are two commonly used techniques of outline analysis (Rohlf 1986; MacLeod 2012a, b). In these methods points (usually a large number) are sampled along the outline but without assigning any homology meaning to them. The coordinates of these points are then used to fit a function, resulting usually in a vector of coefficients of the function. Each such vector mathematically describes its corresponding leaf outline. A matrix of such vectors thus constitutes a dataset expressing the shape variation of a population of outlines, after some appropriate standardization to remove the effects of position, scale and rotation. Outline morphometrics, like landmark techniques, can be used to produce reconstructed shapes, i.e. visually comprehensible results from the analysis of the dataset. Among the various computer implementations of Elliptic Fourier analysis (EFA) are software tools developed by Rohlf (2005), Slice (2008) and Bonhomme et al. (2014).

Other considerations have also played a key part in the success of geometric morphometrics. F.J. Rohlf developed the TPS series of free software tools (e.g. Rohlf 2010a, b, c) which have made geometric morphometric analysis easily available to anyone. His NTSYSpc multivariate analysis package handles these forms of analysis. Rohlf (2014) also created the morphometrics website at the State University of New York, Stony Brook, which remains a key factor in enabling the spread of the new techniques and analytical tools around the world; among much other useful information it includes a glossary of terms used in morphometrics (Slice et al. 2015). Meetings and courses on morphometrics are regularly advertised on the Stony Brook website and there is a comprehensive list of books which map the development of the field and provide in-depth reviews of theoretical issues as well as practical applications.

2.2 Morphometric Analysis of Leaves

Detailed and well-illustrated treatments of leaf descriptors used by taxonomists are available in several recent publications (e.g. Hawthorne and Jongkind 2006; Ellis et al. 2009; Gonçalves and Lorenzi 2011). Typically, leaves vary considerably, even

Fig. 2.1 Variation in leaves taken from a single specimen of *Quercus nigra*

within a single plant, and sampling must take this into account (Fig. 2.1). Hearn (2009), in a study of 2,420 leaves from 151 species, found that an accurate estimate of leaf shape in a species required a minimum of 10 leaves. Blade shape alone is often insufficient to diagnose a species and other leaf features may be needed. In addition, methods of analysis have to be selected according to the problem to be tackled, e.g. when distinguishing leaves of rather similar overall shape, landmark methods are most suitable, but comparison of leaves of very different shape is likely to be more successful using qualitative characters.

2.2.1 Analysis of Conventional Botanical Descriptors

There is a very large literature on the multivariate analysis of taxonomic leaf descriptors, usually in combination with characters from other organs or data types. A good example is the study by Joly and Bruneau (2007) on North American *Rosa* (Rosaceae), which combined morphometric data with cytological and molecular information. Henderson (2011) based his species delimitations of the genus *Geonoma* (Arecaceae) on morphometric analysis of a wide range of taxonomic descriptors, including leaf characters, in an unusually complete implementation of morphometrics as the basis for the species taxonomy in a major monograph.

In most published studies the data was gathered by manual methods, but there are some publications reporting work in which semi-automatic methods of data capture were deployed. West and Noble (1984), White et al. (1988) and McLellan (2000) combined manually guided electronic digitisation of leaf outlines with computerised measurement of conventional leaf descriptors such as length and breadth. Corney et al. (2012) extended this approach by investigating the automatic extraction of taxonomic characters from specimen images, necessitating image processing methods to automatically recognize leaves within images.

2.2.2 Analysis of Leaf Outline Shape

Although variation in leaf outline shape (Fig. 2.2) is traditionally described using a large set of qualitative botanical terms (see e.g. Stearn 1992), many attempts to quantify leaf shape have been made — Melville (1937) and Melville (1951), for example, presented an early manual method using grid coordinates. The development of powerful personal computers and new techniques has made outline analysis more easily available to modern researchers.

Elliptic Fourier analysis (EFA) is the most frequently used technique for quantifying leaf shape (e.g. Jensen et al. 2002; Andrade et al. 2008, 2010; Hearn 2009; Lexer et al. 2009; Chitwood et al. 2014). The digitized outlines of the sampled leaves are decomposed by Fourier analysis into a finite series of ellipses, each represented by four coefficients. Typically, 20–40 harmonics are used, which results in a mathematical description of each leaf as a vector of 80 to 160 numbers (Fig. 2.3). Principal component analysis (PCA) is then usually used to reduce the number of variables and demonstrate the major trends of shape variation in the data set. Extreme shapes along principal components and mean shapes can easily be visualized by reconstruction of shapes from the computed Fourier coefficients (EFDs).

Eigenshape analysis (Ray 1992; MacLeod 1999) is a related technique which uses data obtained from the sequence of angular deviations that occur when progressing along the leaf contour from one digitized point to the next. Principal component variables are then obtained from the data using singular value decomposition. Start and end points of the contours are usually represented by landmarks.

Also related are Contour Signatures, which are sequences of values calculated at successive points along the outline. The centroid-contour distance (CCD) consists of

Fig. 2.2 Examples of leaf shapes

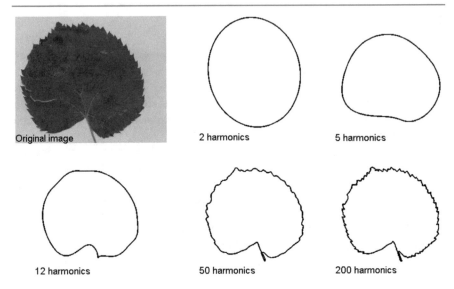

Fig. 2.3 An example of elliptic Fourier analysis. As more harmonics are used to reconstruct the original outline, more detail is preserved

the sequence of distances from the shape's centre to the points along the outline. Other signatures include the sequences of centroid angles and tangents to the contour. The aim is to represent the shape as a numerical vector, independent of the orientation and position of the object within the measuring frame. Various authors have developed refinements of the contour signature approach (Meade and Parnell 2003; Wang et al. 2000, 2003; Ye and Keogh 2009).

Overlap (self-intersection) of leaf parts, especially common in strongly lobed leaves, poses a significant problem for automating the capture of leaf contours. Mokhtarian and Abbasi (2004) and Mokhtarian (1995) developed a method for resolving this difficulty which involved identification of overlap areas using back-lighting and curvature scale space to compare outlines.

Single Parameter Shape Features are quantitative shape descriptors that are intuitive, easy to calculate and of general application. Examples are aspect ratio of leaf blade, ratio of petiole and blade length (Ashby 1948), rectangularity, circularity and perimeter-to-area ratio (McLellan and Endler 1998; McLellan 2005), and invariant moments (Hu 1962; Teague 1980). Methods of analysing such data include the "move median centres" hypersphere classifier (Du et al. 2007; Wang et al. 2008) and neural networks (Wu et al. 2007). Reconstruction of shapes from single parameter shape features is usually not possible and in general the comparison of leaf shapes using only a few such parameters is often unsatisfactory, due to confounding factors such as high correlation between them (McLellan and Endler 1998).

Fig. 2.4 Examples of leaf margins

2.2.3 Analysis of Leaf Margin Patterns

Leaf margins (Fig. 2.4) exhibit a wide morphological variation much used by taxonomic botanists for diagnosing species and genera due to the diverse forms and venation types of the marginal "teeth" (Ellis et al. 2009). As leaf teeth features such as shape, size and number are correlated with climate and growth patterns, they are important for palaeobotanists making inferences about prehistoric climates of fossil species (Royer and Wilf 2006). Hitherto, little use of these features has been made in automated analysis. McLellan and Endler (1998) created an "incision index" as a measure of roughness and Wang et al. (2003) used angular measurements to study marginal variation.

2.2.4 Analysis of Homologous Landmarks and Semi-landmark Configurations

Despite the importance of landmark-based techniques in modern morphometrics in biological research studies (Slice 2005; Weber and Bookstein 2010; Cardini and Loy 2013), relatively few botanical applications have been published and this is probably because most interest has focussed on leaf shapes, which have few good landmarks (Jensen et al. 2002; Volkova and Shipunov 2007; Magrini and Scoppola 2010; Viscosi and Cardini 2011; Klingenberg et al. 2012; Silva et al. 2012; Duminil et al. 2012).

Although landmark methodology has a very well-founded mathematical framework, as mentioned earlier, there are some limitations. Recognition of landmarks from images normally requires manual intervention and this restricts scaling up such studies by automatic extraction. Landmark approaches are also restricted to comparisons between similar leaf shapes, because the technique depends on projecting configurations onto a planar tangent space from non-linear Kendall shape space (Zelditch et al. 2012). Most multivariate statistical methods applied to analysis of landmark configurations use linear combinations of variables, and are only valid for the central part of the tangent projection plane where distortion is at a minimum. Leaf shapes can vary greatly within a single genus (e.g. *Philodendron* in the Araceae) and for these kinds of comparison, landmark methods are unsuitable.

2.2.5   Fractal Dimensions and Polygon Fitting

Fractal dimensions have been used in a few studies for quantitative comparison of leaf shapes (Borkowski 1999; McLellan and Endler 1998; Plotze et al. 2005; Backes and Bruno 2009). Plotze et al.'s study is especially interesting because they compared very different leaf shapes within a single genus, *Passiflora*. However, in general fractal dimension measures alone are probably insufficient to capture and explain enough of the shape variation. Modelling leaves by fitting and comparing polygons has been investigated by Du et al. (2006) and Im et al. (1999).

2.2.6   Analysis of Leaf Venation Patterns

Leaf venation — the pattern of veins at larger and smaller scales — is almost as important in orthodox plant taxonomy as shape for characterizing plant species and genera (Ellis et al. 2009) and of particular importance for leaf fossil identification (Fig. 2.5).

Various methods have been used to capture venation patterns from leaves using image processing methods. Clarke et al. (2006) used scale-space, smoothing and edge detection algorithms. Li et al. (2006) used Independent Component Analysis (ICA, Comon 1994). Mullen et al. (2008) used artificial ants as an edge detection method. Fu and Chi (2006) combined thresholding and neural networks and achieved a good result. Kirchgessner et al. (2002) used b-splines to represent veins and Plotze et al. (2005) used a Fourier high-pass filter combined with a morphological Laplacian operator.

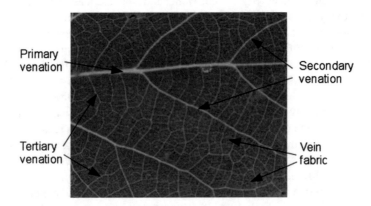

Fig. 2.5 Leaf vein structure

Fig. 2.6 Examples of vein structures

Classification of the extracted vein pattern data has been attempted by Park et al. (2008) using end points and branch points of veins and by Chitwood et al. (2014) using landmark geometrics.

Extraction and comparative analysis of leaf venation patterns continues to be a significant challenge, especially as a technique for large scale implementation. Automated methods are still a long way from realizing the full potential of these features for identification (Fig. 2.6).

2.2.7 Analysis of Leaf Texture

Various studies to capture and compare leaf texture for identification purposes have been published and appear to be most useful when combined with analysis of leaf outline shape. Backes et al. (2009, 2010) used multi-scale fractal dimensions and tourist walks, and Casanova et al. (2009) used Gabor filters. Liu et al. (2009) applied wavelet transforms and support vector machines while Lexer et al. (2009) estimated reflectance from mean greyscale values of leaf images.

2.2.8 Analysis of Other Features of the Leaf Blade

Some preliminary attempts to use 3D imaging and modelling in leaves have been made. Ma et al. (2008) reconstructed leaves and branches from 3D data captured by scanner. Teng et al. (2009) modelled 3D structure and segmentation by combining 2D images, and classified the results into major shape classes. Song et al. (2007) used stereo matching and a self-organizing map to model plant surfaces.

2.3 Morphometrics of Flowers and Other Plant Organs

A number of methods have been proposed to identify plants from digital images of their flowers. Although colour is a more common distinguishing feature here, many methods used to analyze leaf shape can also be deployed. Nilsback and Zisserman

(2007) combined a generic shape model of petals and flowers with a colour-based segmentation algorithm. The end result was a good segmentation of the image, with species identification left for future work.

Das et al. (1999) demonstrated the use of colour alone to identify a range of flowers in a database related to patents covering novel flower hybrids. Their method allows the database to be searched by colour name or by example image, although no shape information is extracted or used. A colour-histogram segmentation method was used by Hong et al. (2004) and then combined with the centroid contour distance (CCD; see Sect. 2.2.2) and angle code histograms to form a classifier. They demonstrated that this method works better than using colour information alone to identify a set of fourteen species. This again suggests that outline shape is an important character to consider, especially in combination with other features.

Elliptical Fourier descriptors (Sect. 2.2.2) were used by Cannon and Manos (2001) on *Lithocarpus* fruits and by Iwata et al. (2004) on vegetative storage organs of *Raphanus*. Yoshioka et al. (2004) and Yoshioka et al. (2007) used EFA to study the shape of the petals of *Primula sieboldii*. Wilkin (1999) used linear measurements of floral organs, seeds and fruits as well as leaves, and principal component methods to investigate whether a closely related group of African *Dioscorea* species were morphologically distinct or not. He discovered that they formed a single morphological entity and hence all belonged to one species. Gage and Wilkin (2008) used Elliptic Fourier analysis on the outlines of tepals (petal-like flower parts) of three closely related species of *Sternbergia* to investigate whether they really formed distinct morphological entities. Clark (2009) used linear measurements of bracts (specialized leaf-like organs associated with the inflorescence) in a study of *Tilia*, and Huang et al. (2006) analyzed bark texture using Gabor filters and radial basis probabilistic neural networks.

Shipunov and Bateman (2005) used analysis of partial warps in a study of the shape of floral labella in *Dactylorhiza*. Gómez et al. (2006) used relative warps analysis to study the evolution of zygomorphy in *Erysimum* flowers. Van der Niet et al. (2010) presented a landmark analysis of zygomorphic floral shape in three dimensions using species of the genus *Satyrium*.

At a smaller scale, the growth of individual grains of barley has been modelled by 3D reconstruction from multiple 2D microscopic images (Gubatz et al. 2007). This allowed both "virtual dissecting" of the grains as an educational tool and also visualization of gene expression via mRNA localization. At a smaller scale still, Oakely and Falcon-Lang used a scanning electron microscope to analyze the vessels found in fossilized wood tissue (Oakley and Falcon-Lang 2009). They used principal component analysis to identify two distinct morphotypes, which correspond to one known and one novel plant species that grew in Europe about 95 million years ago. Butterworth et al. (2009) used sliding semilandmarks to compare the shapes of developing fibres in wild and cultivated cotton. Savriama et al. (2010) studied symmetry and allometry in cells of the alga *Micrasterias* using C.P. Klingenberg's MorphoJ package for geometric morphometric analysis.

A number of studies have used image processing techniques to analyze root structures in the "rhizosphere" (the region in which roots grow, including the soil, soil microorganisms and the roots themselves). For example, Huang et al. (1992) used digital images of roots captured by placing a small camera inside a transparent tube located beneath growing plants. They then used expert knowledge of root shapes and structures (e.g. their elongated shape and symmetrical edges) to combine multiple sources of information and fit polynomial curves to the roots, using a graph theoretic model to describe them. More recently, Zeng et al. (2008) used image intensity to distinguish root pixels from soil pixels and then deployed a point process to combine and connect segments to efficiently identify complete root systems.

These studies show that while the clear majority of botanical morphometrics research publications have focussed on leaves due to their ready availability and usefulness for discriminating taxa, other plant organs, when available, should not be ignored.

2.4   Applications

In this section we move beyond specific algorithms in isolation and methods designed for the laboratory to consider some complete systems and prototypes designed for practical use in the field. In order to have wider impact, it is important to demonstrate that an algorithm can be applied in practice and can be scaled up from a few idealised examples to tackle larger and more complex problems. Here we review systems designed to identify species from plant images, several agricultural applications, scientific research tools for studying species variation and distribution and applications aimed at correlating morphological and climatic variation.

2.4.1   General-Purpose Species Identification

Plant identification is particularly important at the present time because of concerns about climate change and the resultant modifications in the geographic distributions and abundance of species. Development of new crops often depends on the incorporation of genes from wild relatives of existing crops, so it is important to map accurately the distribution of all plant taxa and keep track of their ranges over time. Automated identification of plant species from digital images has become increasingly attractive because of the need to survey the dwindling biodiversity of the world much more rapidly than is possible using traditional taxonomic techniques. Satellite imaging and other such remote surveying technologies can provide useful imagery, but the plant parts thus visualized, e.g. leaves and whole plant shape diagnostics, require novel classification techniques to facilitate accurate identification, given that its ultimate foundation continues to be the orthodox botanical taxonomic system.

As previously discussed in the Introduction, the species to which a plant belongs is an index of critical practical importance. Accurately identifying an organism to its species provides access to existing data linked to the specific name. This information includes such emergent properties as geographical distribution, uses and social or economic value, phytochemistry, breeding potential, functional role in the ecosys-

tems in which it occurs, etc. The binomial botanical name is the label which unlocks
the knowledge about a species that lies scattered throughout the scientific literature.
A robust automated species identification system, when based on expertly identified
training sets, would also encourage and facilitate the participation of a much wider
range of interested people, including those with only limited botanical training and
expertise, to contribute to the survey of the world's biodiversity.

A number of systems have been developed that aim to recognize plant species
from the shapes of their leaves, based on algorithms such as those in Sect. 2.2.2. One
such plant identification system is described by Du et al. (2006). They argue that
any method based on the *global* leaf shape is likely to perform poorly on images
of damaged or overlapping leaves because parts of the leaf perimeter are missing
or obscured. Instead, they suggest that methods based on localized shape features
are more robust for this type of task. Their system matches leaves from images by
fitting polygons to the contour and using a modified Fourier descriptor with dynamic
programming to perform the matching. It aims to be robust with regard to damaged or
overlapping leaves, as well as blurred or noisy images. They claim 92% accuracy for
their method on one sample of over 2000 "clean" images, representing 25 different
species, compared with 75–92% for other methods and that their method is more
robust than others for incomplete leaves or blurred leaf images.

The increasing power and availability of cheap hand-held computers, including
personal digital assistants (PDAs) and smart phones, has led to a number of prototype
applications. A goal widely seen as highly desirable is a portable automated system
that professional botanists and interested amateurs can use to identify plant species
in the field. Although this is a challenging objective, not least because of the very
large number of plant species that may be encountered, there is no doubt as to its
potential importance.

One major ongoing project aims to produce an "electronic field guide" to plants
in the USA (Agarwal et al. 2006). The user photographs a single leaf and the system
will then display images of twenty plant species that have the closest matching
shape according to their Inner-Distance Shape Context algorithm, an approach which
extends the shape context work of Belongie et al. (2002). A related prototype from
the same project includes an "augmented reality" feature (Belhumeur et al. 2008)
and provides a visual display of a herbarium specimen for side-by-side comparison
with the plant in question (White et al. 2006).

The CLOVER system (Nam et al. 2005) allows users to provide a sketch or a
photograph of a leaf using a hand-held computer, which then accesses a remote
server. The server retrieves possible matches based on leaf shape, using several
shape matching methods including an enhanced version of the minimum perimeter
polygons algorithm, and returns the matches to the device to display to the user. The
prototype described is demonstrated to work effectively at recognizing plants from
leaves, using over 1000 images from the Korean flora, with the inevitable trade-off
between recall and precision.

A similar system uses fuzzy logic and the centroid-contour distance to identify
plant species from Taiwan (Cheng et al. 2007). However, this requires the user to

select various characteristics of the plant from a series of menu options, rather than using morphometric analysis directly.

Each of these general-purpose prototypes has been demonstrated to work successfully on at least a small number of species, under more or less stringent conditions. Currently, there is no such system available for everyday use, although interest remains high (Lipske 2008; Kumar et al. 2012).

2.4.2 Agriculture

Rather than trying to identify a plant as belonging to one or other species, it is sometimes sufficient to carry out a binary classification (for example, as healthy or not healthy), without needing to be concerned about the exact taxon to which it belongs. One goal of automated or "precision" agriculture (Burgos-Artizzu et al. 2010) is to allow targeted administration of weed killer, fertilizer or water, as appropriate, from an autonomous robotic tractor, not least to minimize the negative impact on the environment of large scale agriculture. To do this, the system must obviously identify plants as belonging to one category or the other, such as "weed" versus "crop".

As is often the case with machine vision systems, variable lighting conditions can make image processing very difficult. One proposed solution is to control the lighting by building a light-proof "tent" that can be carried on wheels behind a tractor and which contains lamps inside it along with a camera. Such a system successfully distinguished between crop plants (cabbages and carrots) and weed plants (anything else) growing in field conditions (Hemming and Rath 2001). It is questionable whether carrying around such a bulky tent is feasible on a larger scale.

A similar system uses rails to guide a vehicle carrying a camera along carefully laid out plots (Gebhardt et al. 2006). Rather than carrying its own lights, the system is only used under standardized illumination conditions (e.g. bright but overcast). This system extracts shape features such as leaf circularity and area and uses a maximum likelihood estimator to identify leaves that are weeds (specifically dock leaves, *Rumex obtusifolius*) in grassland, with around 85–90% accuracy. A different system to identify dock leaves is described by Šeatović (2008) and uses a scanning laser mounted on a wheeled vehicle to generate 3D point clouds. These are then segmented to separate out leaves from their background and a few simple rules based on leaf size are used to distinguish the dock leaves from other leaves in the meadow.

A related study to distinguish weeds, crops and soil in field conditions uses morphological image processing (Soille 2000). This attempts to identify the centre of each leaf by using colour threshold segmentation and locating the leaf veins. The system finds the veins using a combination of morphological opening and hierarchical clustering. The final classification makes use of a priori knowledge about features of the target plant species. A similar system, combining morphological processing with an artificial neural network classifier, has also been suggested (Pan and He 2008). A combination of colour segmentation and morphological programming has been used to aid the development of a robotic cucumber harvester (Qi et al. 2009). A variety of methods to distinguish various crops from weeds and soil are discussed by Burgos-Artizzu et al. (2010), including colour segmentation and morphological processing. This paper also provides a useful overview of research into "precision agriculture", which aims to use modern technology to optimize crop production, allowing for local variation in soil, landscape, nutrients and so on.

2.4.3 Intraspecific Variation, Geographical Distribution, Climate, Phylogeny

It has long been known that the climate in which a plant grows affects the shape of its leaves (Royer et al. 2005). Recent work has extended this by using digital image analysis to enhance botanical and climatic measurements. Huff et al. (2003) analysed leaves collected from temperate and tropical woodlands and measured the shape factor, finding a correlation with mean annual temperature. The study was then extended to a wider range of environments (seventeen in total) in North America (Royer et al. 2005). A variety of simple digital image analysis methods were used to measure semi-automatically features such as leaf blade area, tooth area and number, major and minor axis lengths. These were then compared to climatic measurements from the different field locations. Finally, correlations between leaf shape and climate were measured, confirming previous findings that plants growing in colder environments tend to have more teeth and larger tooth areas than similar plants growing in warmer environments. One of the goals of this body of work was to support analysis of leaf fossils with the aim of estimating paleoclimatic conditions. By verifying correlations between the shapes of leaves in extant plant species and their environments, it is hoped that fossil leaf shapes can be used to hypothesize how the Earth's climate has changed in the past, at both global and local scales.

An early study by Dickinson et al. (1987) used manual digitization with a tablet to identify landmarks on leaf cross-sections and principal component analysis to analyse the resulting data. They identified geographically correlated variation between collection sites and identified intermediate forms among the specimens, suggesting the occurrence of hybridization. Work by Wilkin (1999) and Gage and Wilkin (2008) used morphometric analysis to determine species boundaries, a goal just as important as identifying the species to which a particular specimen belongs.

One of the most significant developments in comparative biology in the last 30 years has been the development of phylogenetic reconstruction methods. Phylogenetic or cladistic analysis, as this research field is called, is today dominated by the use of nucleic acid or protein molecular data, but analyses using morphology remain important, both for comparison with molecular results and because of the need to include fossils in classifications.

Phylogenetic analysis has been used mostly for establishing evolutionary relationships between taxa above the species level — higher taxonomy — while morphometric approaches have proved to be most useful at species level and below. This is reflected in the fact that phylogenetic systematics tends to use discrete (usually qualitative) characters, while morphometric studies predominantly focus on quantitative (real number-valued) variables.

Another difference is that cladistic analysis uses the tree paradigm and models the phylogeny of the organisms by searching the data set for optimal nested arrangements of discrete character values on the internal branches (synapomorphies). The resulting cladogram is an hypothesis of the phylogenetic history of the study organisms and depends only on the characters which have been selected by the analysis as synapomorphies. In contrast, the methods of morphometrics are more appropriate for dealing with character variation — a major confounding factor in identification of species and infraspecific taxa. Tree methods such as hierarchical cluster analysis and classification and regression trees are used, but ordination is also important. The character data is often synthesized into new variables such as principal components or distance parameters which can obscure the relative importance of different variables in the mathematical criterion ultimately used to distinguish groups of individuals. Morphometrics freely transforms original variables into a wide variety of derived mathematical forms (e.g. z-values, logarithms, square roots, etc.), while phylogenetic studies tend to use techniques which keep the characters in their original form, or merely transform their values into probabilities.

2.5 Summary

In this chapter, various morphometric methods used hitherto in botanical studies have been mentioned and the historical development of the subject has been briefly surveyed. Because of the extensive range of problems to which morphometrics has been applied, a diversity of analytical techniques has become available over the years. Even within the restricted area of using morphometrics for plant identification — the subject of this book — appropriate methods must be chosen for the task at hand. Plants are extremely diverse in structure, shape, size and colour. A method that works well in one plant group may rely on features that are absent in another, e.g. landmarks may be readily definable for species with distinctively lobed leaves, but not for those with simple unlobed leaf blades.

The very large number of plant species and the morphological variation usually found within any one of them, means that the development of automated identification systems is ultimately a large scale enterprise, requiring processing of high numbers of digital images. Automation is therefore essential since any system that requires significant manual effort, for example in tracing leaf outlines or accurately locating landmarks on images, is unlikely to be practical when scaled up to thousands of specimens. However, there will be contexts in which the user needs to be directly involved at some critical stage in the identification process: if an electronic field-guide provides e.g. ten predictions of species, rather than just one, the user will need to make the final choice (Agarwal et al. 2006). A related issue is how fast processing speed needs to be. The user of a hand-held electronic field-guide is likely to require responses interactively and so (near) instantaneously, whereas for an identification tool to be used on a large set of images in a herbarium or laboratory, it may be acceptable to wait overnight for a comprehensive response — assuming no human interaction is needed in this case.

Feature Extraction

3

> **Chapter contributions:**
>
> • Comparative study of leaf-shape analysis techniques.
> • Methods for extraction and description of leaf macro- and micro-texture, margin characters, and venation patterns.

As discussed in Chap. 2, plant leaves contain many different morphological components (characters) that can be used in analysis for identification, such as outline shape, margin details and venation type. It is important to extract these components accurately from the leaves and generate appropriate descriptors for them.

Most previous work in this field has focussed only on outline leaf shape, using a wide array of traditional and leaf-specific shape analysis techniques. This chapter begins with a comparative study of the most popular shape-based methods, before continuing on to the primary focus, which is to present new methods based on other major leaf components, namely texture, margin and venation.

3.1 Leaf Shape

There are several reasons for the emphasis on leaf outline shape in previous studies. In the first place, it is perhaps the most obvious aspect to use - the wide variety of shapes is easily observed and might suggest a high discriminative power. The outline is also particularly easy to extract and can be done by simple thresholding if the leaf is set against a plain background. There is moreover a large array of existing morphometric and shape analysis techniques that can be applied to the problem. In this section, the most commonly used techniques will be applied to the same dataset, in order to evaluate the advantages of each.

© Springer-Verlag GmbH Germany 2017
P. Remagnino et al., *Computational Botany*, DOI 10.1007/978-3-662-53745-9_3

3.1.1 Study of Existing Techniques for Leaf Shape Analysis

Most of the methods that have been used hitherto for analysing leaf shape are unsuitable for automatic leaf classification. Linear measurements, such as angles and the length of segments measured between certain recognized structural points on the leaf (Haigh et al. 2005), have a natural appeal to botanists due to their similarity or indeed congruence with traditional botanical descriptors. However, some previous categorization of the leaves is often required, for example as lobed, unlobed or palmate (lobes radiating from the base of the leaf), in order to select an appropriate set of measurements and manual intervention is needed to locate the correct measuring points. An example is the study of Jensen et al. (2002) which compared leaves of two 5-lobed species of *Acer* (maples) using the relative positions of the lobe tips and sinus bases. The use of landmarks tends to be restricted to studies involving comparison of species with similar leaf shapes, due to the lack of landmarks that can be said to be common across all types of leaves and also to the limitations of the tangent shape space projection from full Kendall shape space inherent in modern geometric morphometric approaches (Zelditch et al. 2012).

In the following evaluation, three common and widely applicable methods will be used: shape features, centroid-contour signatures and elliptic Fourier descriptors. Whilst several other techniques have been tried, including fractal dimensions and polygon fitting, their use has been limited and they do not appear to have any advantages over the three techniques we select here.

3.1.1.1 Shape Features (SF)

Shape features are sets of geometrical and morphological characteristics, not specific to leaves, selected as appropriate for adequately describing a leaf. They are appealing since it is immediately apparent how a feature such as the aspect-ratio communicates intuitively one characteristic of a leaf's shape. However, shape features contain insufficient information to permit a precise reconstruction of the leaf shape and it is quite possible that two visibly different leaves could produce the same set of values of the features used. A further problem is the likelihood of high levels of correlation between some features, although this can be resolved through the use of feature selection techniques. While there is no definitive set, certain shape features suitable for leaf classification have frequently been used and the set used here includes many of these.

First, the minimum bounding rectangle (MBR) and convex hull (CH) of the leaf are calculated. The leaf is oriented through calculation of the MBR. The following variables can then be defined: height, h, and width, w, of the MBR; area of the leaf, the MBR and the CH respectively as A_L, A_{MBR} and A_{CH}; perimeter of the leaf, the MBR and the CH respectively as P_L, P_{MBR} and P_{CH}; the minimum and maximum distances from the centroid to the contour, CCD_{min} and CCD_{max}. The eight shape features are then constructed from these variables as follows:

1. Aspect ratio

$$F_1 = \frac{h}{w}$$

2. Rectangularity

$$F_2 = \frac{A_L}{A_{MBR}}$$

3. Ellipticality

$$F_3 = \frac{4A_L}{hw\pi}$$

4. Solidity

$$F_4 = \frac{A_L}{A_{CH}}$$

5. Perimeter Convexity

$$F_5 = \frac{P_L}{P_{CH}}$$

6. Sphericity

$$F_6 = \frac{CCD_{min}}{CCD_{max}}$$

7. Form factor

$$F_7 = \frac{4\pi A_L}{P_L}$$

8. Gravity

$$F_8 = \left| \frac{\bar{y}}{h} - \frac{1}{2} \right|$$

3.1.1.2   Centroid-Contour Signatures (CCS)

Contour signatures are sequences of values calculated at points spaced along the perimeter of a shape. Among those that have been defined, the most commonly used is the centroid-contour distance — the sequence of distances from the centroid of the shape to each boundary point. Points on the contour may be evenly spaced either in terms of their distance along the perimeter or as the result of equi-angular displacements around the centroid. Problems arise with this latter method when a line extended from the centroid crosses the contour more than once. Some authors, like Meade and Parnell (2003), have used uneven spacing by increasing point sampling in regions of higher curvature of the contour.

Here we use two different signatures, the centroid-contour distance (CCD) and the centroid angle signature (CAS).

$$CCD(i) = \sqrt{(x_i - x_c)^2 + (y_i - y_c)^2}$$

$$CAS(i) = \left| \tan\left(\frac{x_i - x_c}{y_i - y_c} \right) - \tan\left(\frac{x_0 - x_c}{y_0 - y_c} \right) \right|$$

where x_i, y_i are the x and y co-ordinates respectively for the ith contour point, and x_c, y_c are the x and y co-ordinates of the leaf centroid.

The CCS is normalised so that all values in the sequence sum to unity. When comparing two pairs of signatures, orientation invariance is achieved via cross-correlation, whereby the distance between them is measured for every offset of one against the other, and the minimum of these distances is used. This is equivalent to rotating one leaf in relation to another until the difference between the two is minimised.

3.1.1.3 Elliptic Fourier Descriptors (EFD)

The elliptic Fourier descriptor of a shape consists of the set of coefficients for the first k harmonics of the elliptic Fourier expansion of the contour coordinates. These are given, for the nth harmonic, as

$$a_n = \frac{P}{2n^2\pi^2} \sum_i \frac{\Delta x_i}{\Delta p_i} \left(\cos \frac{2n\pi p_i}{P} - \cos \frac{2n\pi p_{i-1}}{P} \right)$$

$$b_n = \frac{P}{2n^2\pi^2} \sum_i \frac{\Delta x_i}{\Delta p_i} \left(\sin \frac{2n\pi p_i}{P} - \sin \frac{2n\pi p_{i-1}}{P} \right)$$

$$c_n = \frac{P}{2n^2\pi^2} \sum_i \frac{\Delta y_i}{\Delta p_i} \left(\cos \frac{2n\pi p_i}{P} - \cos \frac{2n\pi p_{i-1}}{P} \right)$$

$$d_n = \frac{P}{2n^2\pi^2} \sum_i \frac{\Delta y_i}{\Delta p_i} \left(\sin \frac{2n\pi p_i}{P} - \sin \frac{2n\pi p_{i-1}}{P} \right)$$

where x_i, y_i are the coordinates for the ith point, p_i is the distance along the contour to the ith point, P is the total perimeter distance, and Δx_i, Δy_i, Δp_i are the respective distances between points i and $i-1$.

$$\Delta x_i = x_i - x_{i-1}$$

$$\Delta y_i = y_i - y_{i-1}$$

$$\Delta p_i = p_i - p_{i-1}$$

Elliptic Fourier descriptors are popular with botanists because the leaf shape can easily be reconstructed from the descriptor. The main trends of variation within a dataset revealed by principal component analysis can be visualised from elliptic Fourier descriptors that increase or decrease along each principal component (Andrade et al. 2008; Furuta et al. 1995; Lexer et al. 2009; Yoshioka et al. 2004).

3.1.2 Evaluation and Results

The methods were evaluated on a dataset containing sixteen leaves from each of 100 different species. A 16-fold cross-validation was performed, such that one leaf from each species was used each time as the testing set, whilst the remaining leaves were used as the training set. Classification was performed using the k-nearest-neighbour technique, with $k = 15$. Table 3.1 shows the average rate of correct classification and standard deviation for each method. The standard deviation given here is the deviation in classification rates between different species.

As can be seen, all three methods performed similarly with the shape features producing slightly better results at 60.8% correct classification. The shape features also had a lower standard deviation than the other two methods, at 26.3%, although all three had quite high values.

This high variance can partly be explained by examining the species on which each method performed best. Whilst the CCSs and EFDs performed similarly on most species, there were some on which they both performed significantly better than SFs, while in some others it was the SFs which performed best. Examples can be seen in Figs. 3.1 and 3.2 respectively. Species for which SFs performed best typically had more complex outlines and often high levels of intra-species variation. Many had lobed leaves with varying numbers of lobes from leaf to leaf, whereas species for which the number and position of lobes on each leaf remained constant tended to achieve comparable results for all methods. In contrast, CCSs and EFDs achieved better results on leaves with simpler non-lobed shapes and lower intra-species and inter-species variation. The explanation appears to be that SFs capture the general properties of the shape and so are more resilient to slight variations,

Table 3.1 Cross-validation results for the three shape analysis methods

Method	Correctly classified (%)	Standard deviation (SD) (%)
Shape features (SF)	60.8	26.3
Contour signatures (CCS)	59.6	33.4
Elliptic fourier descriptors (EFD)	57.8	30.7

Fig. 3.1 Examples of species for which elliptic Fourier descriptors and contour signatures are more likely to achieve correct classification than shape features

Fig. 3.2 Examples of species for which shape features are more likely to achieve correct classification than elliptic Fourier descriptors or contour signatures

Fig. 3.3 Species retrieval results showing the proportion of cases (y-axis) for which the correct species is returned within the top n (x-axis), for shape features (SF), centroid-contour signatures (CCS) and elliptic Fourier descriptors (EFD)

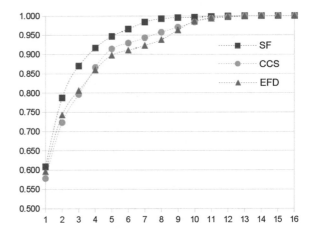

whereas CCSs and EFDs capture more precisely the subtler details required for distinguishing between species with similar-shaped leaves.

Although the accuracy of all three methods was relatively low on a dataset of this size (100 species) when formulated as a retrieval problem where the top n species are returned, shape-based methods can be seen as an effective means of eliminating the majority of species and thus improving the accuracy and reducing the computational time required by other methods performed subsequently.

In Fig. 3.3, the y-axis indicates the proportion of cases for which the correct species is included within the top n results returned, where n is the value on the x-axis; e.g. approximately 90% of leaves tested using EFDs included the correct species within the first five species returned. For SFs, the correct species is included in the first eight species returned in over 99% of cases, whilst for all methods the correct species always appears within the first 14 out of the total 100 species. This result will be explored further in Chap. 4.

3.2 Leaf Texture

Much of the texture present on a leaf is due to its venation, but other sources include hairs and glands. Leaf veins typically have a hierarchical structure and can be separated into two main groups: the lower order vein framework, consisting of the larger primary and secondary veins, and the higher order vein fabric which occupies the spaces in between (see Figs. 2.5, 3.4). Because of this, it is beneficial to analyse the texture at both macro and micro scales, using descriptors that capture, respectively, features of the entire surface of the leaf or only of areas between the larger veins.

Another aspect of leaf texture to be considered is the difference between the lower (abaxial) and upper (adaxial) sides. Typically, the venation is more prominent on the abaxial side, with hairs and glands also being more common features there, whilst the adaxial side is more likely to have a waxy texture. These differences are related to essential functional requirements — the leaf must restrict water loss through the numerous pores (stomata) that cover its surface but simultaneously maximize the gas exchange with the atmosphere, on which photosynthesis depends.

3.2.1 Macro-texture

Here, the objective is to generate a descriptor for the entire surface of the leaf that will take account of variation in such factors as lighting conditions or damage caused by insects or disease.

For each leaf image, a large number (up to 1024) of small, fixed-size windows are randomly selected from the surface of the leaf. For each window 20 features are calculated based on the responses from different filters applied to all the pixels in the window.

The filters used are a rotationally invariant version of the Gabor filter:

$$g(x, y) = \exp \frac{r^2}{2\sigma^2} \cos \frac{2\pi r}{\lambda}$$

where $r = \sqrt{x^2 + y^2}$ is the distance from the centre of the filter, σ is the standard deviation, and λ is the wavelength, set to be $\lambda = 3\sigma$. Five differently scaled filters are used, produced by varying σ. The wavelength is fixed in relation to σ so that the filters are scaled versions of each other. Each filter is convolved with the window and four features are then calculated for that filter for the window:

1. Average positive value

$$\sum_{\substack{(i,j) \in W \\ s_j \geq 0}} \frac{f_{ij}}{|W|}$$

2. Average negative value

$$\sum_{\substack{(i,j) \in W \\ s_j \leq 0}} \frac{f_{ij}}{|W|}$$

3. Energy

$$\sum_{(i,j) \in W} \frac{f_{ij}^2}{|W|}$$

4. Entropy

$$- \sum_{(i,j)\in W} \frac{|f_{ij}|}{|W|} \log \frac{|f_{ij}|}{|W|}$$

Where W is the current window, f_{ij} is the response for the current filter at pixel (i, j), and $|W|$ is the size of the window.

Further details and analysis are presented in Sect. 4.1.1 including discussion of the parameters used and techniques for classifying leaves based on the data generated.

3.2.2 Micro-texture

When working at a high enough resolution to allow extraction of useful information from the finer vein fabric, it is not practical to cover the entire leaf surface, due to computational requirements. Instead, a selection of sample windows can be chosen. If texture samples (windows) are extracted randomly from a leaf, the level and quality of the vein framework present in a sample may vary greatly, depending on the precise position of each window on the leaf. A simple method is therefore suggested for extracting samples that as far as possible contain only the vein fabric, which should result in the contents of these samples being more consistent (Fig. 3.4).

The first stage is to reduce the scale of the image by convolving it with a Gaussian kernel and then sub-sampling. This has the effect of smoothing out much of the detail in the vein fabric, whilst retaining the main venation. Next, the image background – the paper on which the leaf is mounted – is removed. This can be done using Otsu's thresholding method (Otsu 1979). An edge detection operator is then applied to the foreground of the image to provide a rough measure of the areas with strong edges in this scale space. A large number of potential windows (10 times the final number to be used for analysis) are sampled at random from the foreground, which consists

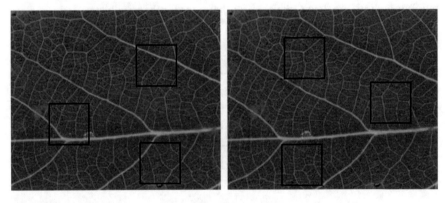

Fig. 3.4 Random sampling (*left*) compared with desired sampling (*right*), shown on an x-ray image for increased contrast

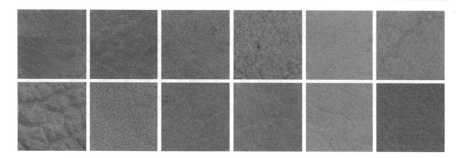

Fig. 3.5 Extracted texture samples from twelve species of *Quercus* (Oak)

only of the leaf, and are sorted according to the sum of the squared edge magnitude for all the pixels within the window. The desired number (in this case, eight) of non-overlapping sample windows with the lowest sum can then be selected for use. Examples of windows selected by this method are shown in Fig. 3.5.

3.2.2.1   Gabor Filters

The texture analysis method presented here is based on the joint distribution of responses to Gabor filters. A Gabor filter (Daugman 1985) is essentially a sinusoid modulated by a Gaussian function. It can be expressed as follows:

$$G(x, y) = \exp\left(\frac{x'^2 + \gamma^2 y'^2}{2\sigma^2}\right) \cos\left(\frac{2\pi x'}{\lambda} + \psi\right), \qquad (3.1)$$

where:

- $x' = x \cos\theta + y \sin\theta$
- $y' = y \cos\theta - x \sin\theta$
- θ is the orientation of the filter.
- γ is the filter aspect ratio.
- σ is the standard deviation of the Gaussian.
- λ is the wavelength of the sinusoid.
- ψ is the phase offset.

Gabor filters have been applied to a large range of computer vision problems including image segmentation (Sandler and Lindenbaum 2006) and face detection (Huang et al. 2005). Of particular interest are the links found between Gabor filters and the human visual system (Daugman 1980) (Fig. 3.6).

3.2.2.2   Texture Analysis from Gabor Co-occurrences

A bank of 128 Gabor filters is created, where for filter G_{mn}, $\sigma = 1.5 * 1.2^{m-1}$, $\lambda = \frac{\sigma\pi}{2}$ and $\theta = \frac{n\pi}{16}$, with $m \in [0, 7]$ and $n \in [0, 15]$ referring to the filter scale

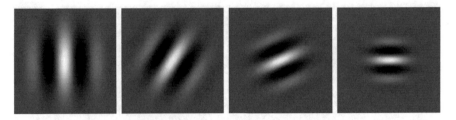

Fig. 3.6 Examples of the Gabor filters used here

Table 3.2 Examples of Gabor co-occurrence matrices for three species of *Quercus*

	C_{01}	C_{02}	C_{03}	C_{04}	C_{05}	C_{06}	C_{07}

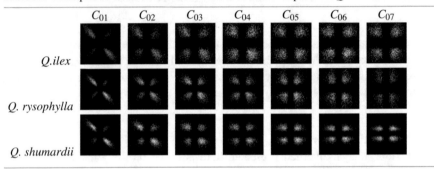

Q.ilex

Q. rysophylla

Q. shumardii

and angle respectively. For all filters, $\gamma = 1$ and $\psi = 0$. The full set of filters is applied to each texture, but for each scale only the value corresponding to the highest absolute response for all the orientations is recorded for each pixel. This ensures that the method is rotation invariant.

The results of the filtering for an image are combined into a series of co-occurrence matrices (Haralick et al. 1973), whereby for each pair of scales, the resulting matrix describes the probability of a pixel producing one response value for the first scale, and another for the second.

$$C_{kl}(i, j) = P(g_k(x, y) = i, g_l(x, y) = j) \tag{3.2}$$

where $g_m(x, y) = \max_{n=0...15} (G_{mn}(x, y) * I(x, y))$ is the maximum response from convolving the filters for scale m with the image I at point (x, y), and (i, j) is a pair of response values. Examples of these matrices for values $k = 0$ and $l = 1$ to 7 are given in Table 3.2. The x-axis and y-axis for each matrix cover the range of response values for each of the two filters, with the greyscale value representing the frequency at which the two filters gave the corresponding pair of responses.

3.2.2.3   Classifying Micro-textures

To classify textures, the corresponding co-occurrence matrices for different textures are compared directly. This is done by treating the co-occurrence matrices as probability distribution functions (pdfs), by simply dividing each value by the sum of all values, and using the Jeffery-divergence distance measure – the symmetric version of the Kullback-Leiber divergence (Budka et al. 2011). For two pdfs, f_a and f_b, the distance between them, $JD(f_a, f_b)$, is calculated as follows:

$$JD(f_a, f_b) = \sum_i \sum_j f_a(i,j) log \frac{2f_a(i,j)}{f_a(i,j) + f_b(i,j)} + f_b(i,j) log \frac{2f_b(i,j)}{f_a(i,j) + f_b(i,j)} \qquad (3.3)$$

The distance between two images A and B is then:

$$D(A, B) = \sum_k \sum_{l,l \neq k} JD(C_{kl}^A, C_{kl}^B) \qquad (3.4)$$

where C_{kl}^A and C_{kl}^B are respectively the co-occurrence matrices at scale k, l for images A and B.

The final classification is performed using the k-nearest neighbours method, with $k = 3$. The most frequent class of the three closest texture samples to the one being classified is chosen. In the case that all three classes are different, the class of the single closest texture sample is used instead. This strategy was chosen as it reduces the risk of classification errors due to outliers.

3.2.2.4   Experiments

Datasets

The method was evaluated using two texture datasets. The first was constructed using the method described previously. For each of eight leaves from 32 different species, eight 64×64 windows were selected. This window size was chosen as it was found to be small enough to allow the windows to fit between the main veins in leaves with dense vein fabrics. Eight windows were then used to provide an adequate overall sample size, because more would require more computation and might be too many to fit on particularly small leaves without significant overlap between windows. Each of the eight samples for a leaf was filtered using the set of Gabor filters, before being combined into a single set of co-occurrence matrices. The second dataset used eight windows sampled at random from the same leaves to illustrate the value of the texture extraction method for selecting suitable windows.

Comparison Methods

For comparison, the above datasets were also used with a number of traditional texture analysis methods:

- Fourier Coefficients:
  The Fourier Transform of each window was calculated. From this, a vector of 64 features was found, whereby the ith feature $f_i = \sum_{\theta=0}^{\pi} |F(i\frac{w}{64}, \theta)|$, where $F(r, \theta)$ is the Fourier Transform in polar form, and w is half the image width (Journaux et al. 2008).

Table 3.3 Results for the two datasets (% correctly classified)

	Vein fabric	Random windows
Gabor co-occurrences	85.16	79.69
Fourier	82.42	62.89
Gabor	50.78	45.70
Co-occurrence matrices	69.14	61.72

- Gabor Filters:
  The same set of Gabor filters used in Sect. 3.2.2.2 is applied to each image. The energy in each resulting image is then calculated as $e_{\sigma\theta} = \sum_x \sum_y (G_{mn}(x, y) * I(x, y))^2$. The set of energies for each scale are then averaged, resulting in eight rotationally invariant features.
- Co-occurrence Matrices:
  The traditional co-occurrence matrices were produced, using angles of $0\ rad$, $\frac{\pi}{4}\ rad$, $\frac{\pi}{2}\ rad$ and $\frac{3\pi}{4}\ rad$ and distances of 1, 2 and 3. For each distance, a set of 14 textural features was calculated, as described by Haralick et al. (1973).

3.2.2.5   Results

The results for the experiments are given in Table 3.3, with values representing the percentage of leaves correctly classified. All the algorithms performed better on the dataset created using the method of Sect. 3.2.2 than on the dataset of randomly selected windows, showing the value of this method of leaf texture extraction. For all datasets the new method performed best, with the basic Gabor method performing worst. The improvement between the two datasets was greatest for the Fourier method, suggesting that it is better at capturing the finer detail of the vein fabric; however its performance was still not able to match the proposed method.

3.3   Margin Characteristics

The leaf margin varies greatly between species in its detailed form – particularly in the presence and shape of marginal teeth – and this can provide valuable information for species identification as well as about the environment in which the plant occurs (Fig. 2.4). In the case of leaf fossils this is of particular interest as an indicator of prehistoric climates and climate change. Extracting the detailed form of the leaf margin accurately and independently of the overall leaf shape is therefore important to ensure meaningful comparison of the margins of different leaves.

3.3.1 Extracting the Margin

The first step is to extract the leaf margin. Having extracted a mask of the leaf, a modal filter is applied to the mask to acquire a smoothed version of the leaf shape (see Fig. 3.7). This filter sets each pixel of the smoothed leaf to be part of the leaf if the majority of the original pixels within the filter's radius were part of the leaf, and to be part of the background otherwise. This has the effect of removing the tips of the teeth and filling in any small gaps. By varying the filter's radius different levels of smoothing can be achieved and with sufficiently large radii all detail can be removed from the margin. In this case a radius of 15 pixels was used. From this smoothed outline, m evenly spaced points around the contour are calculated (in this case, $m = 8192$ was used), encompassing the entire outline. For each of these points a corresponding point on the original outline is then calculated. This is done by first estimating the line perpendicular to that which joins the two points lying at distance k on either side of the current point. Then the sub-pixel point at which this perpendicular intersects the original leaf outline is found by linear interpolation. The distance between this point and the current point is then calculated (see Fig. 3.8), and these distances for all the points in the smoothed outline are combined to produce a margin signature, $\mathbf{s} = \langle s_1, \ldots, s_m \rangle$. Examples of extracted margin signatures are shown in Fig. 3.9.

The extracted margin is partitioned into n overlapping windows, $\mathbf{x} = \langle x_1, \ldots, x_n \rangle$, of equal size and spacing (in this case $n = \frac{m}{8}$ and the window size used is $\frac{m}{128}$). This is done for a number of reasons. First, the exact number of teeth will vary between leaves of the same species, making alignment of their margins problematic. Using windows is more robust to this variation because each window describes the margin segment within it. This also provides an adequate description of the margin while using a much smaller number of datapoints, thus reducing computation time by a factor of one-eighth. By overlapping the windows, sensitivity to their exact position in $\mathbf{s}$ is reduced.

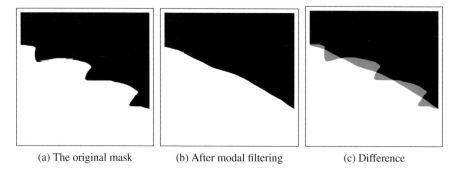

(a) The original mask   (b) After modal filtering   (c) Difference

Fig. 3.7 Using modal filters to extract the leaf teeth

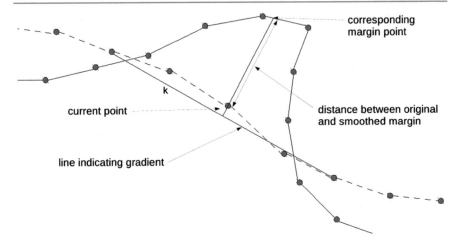

Fig. 3.8 Calculating a point on the margin signature

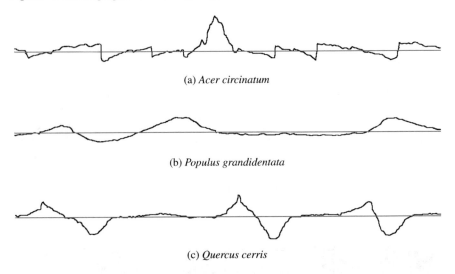

(a) *Acer circinatum*

(b) *Populus grandidentata*

(c) *Quercus cerris*

Fig. 3.9 Examples of segments from extracted margin signatures. The y-axis represents the distance between the smoothed margin and the original margin

For each point within a window, x_i, three values are calculated:

1. Magnitude - This is the signed distance between the smoothed margin point and its corresponding point on the original margin; the sign is determined by whether the original margin point lies inside or outside of the smoothed margin.
2. Gradient - The signed difference between the current point in the margin signature and the next point.

3. Curvature - The angle at the current point between the previous point and the next
   point in the signature.

For each of these, two features are then calculated for the window, giving a total
of 6 features per window:

- Average positive value:
$$\sum_{\substack{s_j \in x_i \\ s_j \geq 0}} \frac{s_j}{|x_i|}$$

- Average negative value:
$$\sum_{\substack{s_j \in x_i \\ s_j \leq 0}} \frac{s_j}{|x_i|}$$

Where x_i is the current window, s_j is the value at a point within the signature, and
$|x_i|$ is the size of the window.

3.4 Locating the Apex and Insertion Point

Among the great diversity of leaf shapes (Fig. 2.2), the only two landmarks which
are consistently present are the 'insertion point' - where the petiole (leaf stalk), joins
the leaf - and the apex (leaf tip). It is therefore important to locate these two points
and the extracted margins can be utilised to do this.

Potential candidate points are first identified by selecting the local maxima from
the margin signature which have an absolute magnitude greater than 25% of the
global maximum. The value of 25% was used as it was found to be small enough
that for all leaves in the dataset the true apex and insertion points were always among
the candidates selected. Based on the principle that each side of a leaf - from insertion
point to apex - is an approximate reflected image of the other, dynamic time warping
can be used to identify the two points on each margin which differ least between the
two sides and so are most likely to be the insertion point and apex.

3.4.1 Dynamic Time Warping

Dynamic time warping (henceforth DTW, Sakoe and Chiba 1978) is a technique for
measuring the similarity between two different sequences. During the comparison, it
allows parts of the signals to be stretched or compressed to a certain extent, thereby
accounting for the sequences being of different lengths (for instance, due to differences in speed) or containing natural distortions. A typical application of DTW is
speech recognition, where people may speak at different speeds, or elongate different
sounds. It has also been used for a number of computer vision problems, including

face detection (Turkan et al. May 12, 2006) and action recognition (Sempena et al. 2011).

Given two sequences $\mathbf{x} = \langle x_1, \ldots, x_m \rangle$, $\mathbf{y} = \langle y_1, \ldots, y_n \rangle$, an $m \times n$ cost matrix $\mathbf{C}$ is calculated, whereby value c_{ij} is the distance between points x_i and y_j. Under the assumption that point x_1 corresponds to point y_1 (i.e. the same starting points), and x_m to y_n, a monotonic path through $\mathbf{C}$ is found, beginning at c_{00} and ending at c_{mn}, such that the sum of the values at the nodes visited is minimized. This path then represents the optimal alignment of points in $\mathbf{x}$ to those in $\mathbf{y}$. This can be calculated relatively efficiently (quadratic complexity) by recursively accumulating the costs in a matrix $\mathbf{D}$. The value d_{ij} is calculated as follows:

$$
d_{ij} = \begin{cases}
c_{ij} & \text{if } (i = 1 \wedge j = 1) \\
\infty & \text{if } (i = 1 \wedge j > 1) \vee (j = 1 \wedge i > 1) \\
c_{ij} + \min \begin{pmatrix} d_{i-1,j} \\ d_{i,j-1} \\ d_{i-1,j-1} \end{pmatrix} & \text{otherwise}
\end{cases}
\tag{3.5}
$$

Once all the values in $\mathbf{D}$ have been calculated, the measure of the similarity between the two sequences is given by d_{mn}.

There are a number of extensions to the standard DTW algorithm that have been proposed in the literature (Sakoe and Chiba 1978; Salvador and Chan 2007). Calculating d_{ij} by using Eq. (3.5) results in a path which travels monotonically between adjacent cells, either horizontally, vertically or diagonally. Since a continued horizontal or vertical movement represents the compression of a subsequence to unit length, or the stretching of a single point to a much longer length, this could result in unrealistic distortions. To counter this we add the condition that every step that is made horizontally or vertically must also be accompanied by a diagonal step (see Fig. 3.10). This restricts the maximum distortion of a subsequence to a level that is realistic for this type of data, whilst ensuring that distortions carry an additional cost because they produce longer paths.

If two sequences are similar, the optimal DTW path will be close to the diagonal of the cost matrix, where $i = j$. If the optimal path diverges from this by more than a certain amount it is unlikely that the two sequences are from the same class. This allows a constraint to be added to improve the speed of the algorithm. By only calculating d_{ij} for $|i - j| \leq k$, the complexity can be reduced from $\mathcal{O}(n^2)$ to $\mathcal{O}(kn)$ where $k \ll n$, without the risk that this would produce a sub-optimal path if the two sequences were from the same class (Salvador and Chan 2007). With these improvements included, the equation for calculating d_{ij} becomes:

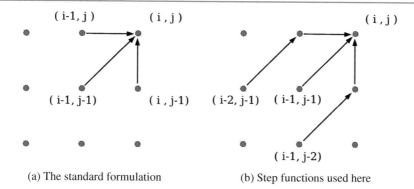

(a) The standard formulation   (b) Step functions used here

Fig. 3.10 Legal steps for a path when calculating DTW

$$
d_{ij} = \begin{cases}
\infty & \text{if } |i - j| > k \\
c_{ij} & \text{if } (i = 1 \land j = 1) \\
\infty & \text{if } (i = 1 \land j > 1) \lor (j = 1 \land i > 1) \\
c_{ij} + \min \begin{pmatrix} d_{i-2,j-1} + c_{i-1,j}, \\ d_{i-1,j-2} + c_{i,j-1}, \\ d_{i-1,j-1} \end{pmatrix} & \text{otherwise}
\end{cases}
$$

$$(3.6)$$

3.4.2 Finding the Points of Margin Symmetry

For a given candidate point, the corresponding window x_i is identified from the circular sequence of windows $\mathbf{x} = \langle x_1, \ldots, x_n \rangle$ for the leaf. Two sequences, $\mathbf{a} = \langle a_1, \ldots, a_{\frac{n}{2}+w} \rangle$, $\mathbf{b} = \langle b_1, \ldots, b_{\frac{n}{2}+w} \rangle$ are generated, where $a_1 = b_1 = x_i$, $a_j = x_{i+j}$, $b_k = x_{i-k}$ and n is the total number of windows. Since the insertion point and apex may not lie directly opposite each other, the sequences $\mathbf{a}$ and $\mathbf{b}$ are continued for a distance of w beyond the mid point $x_{i+\frac{n}{2}}$, so that the ends of the sequences overlap. A value of $w = \frac{n}{8}$ was used.

The accumulated cost matrix $\mathbf{D}$ is generated as described in Sect. 3.4.1. Because the last w points in the two sequences are the reverse of each other, similarity is calculated as the minimum d_{jk} where $j + k = n$. Using this method, the insertion point and apex were correctly identified in 97.75% of test cases with one or the other being correctly found in 99.25% of the 1600 leaves in the dataset.

These apices and insertion points are used in Sect. 4.1.2 for carrying out leaf classification based on the leaf margin descriptors described above.

3.5 Venation Patterns

In this section two methods for the extraction of leaf venation are presented. The first uses a genetic algorithm to evolve classifiers for detecting veins on a pixel-by-pixel basis while the second utilizes an ant colony algorithm to attempt the extraction of continuous vein segments.

3.5.1 Extraction by Evolved Vein Classifiers

3.5.1.1 Classifying the Vein Pixels
In this procedure a genetic algorithm is used to evolve a set of classifiers for detecting vein pixels. Each classifier consists of a pair of bounds for each of the features used. If the values of all the features for a pixel fall within all the bounds for a classifier, then it is classified as vein. The vein pixels found by all the classifiers in the set are combined, and all other pixels are classified as non-vein. These classifiers are similar to those used by Liu and Tang (1999). More specifically, the set of vein pixels, V, is determined as follows:

$$V = \{(x, y)|0 \leq x < w, 0 \leq y < h,$$
$$\exists c \in C | \forall f_i \in F_{xy}, c_{i0} \leq f_i \leq c_{i1}\}$$

where

- w, h are the image height and width respectively
- C is the set of all classifiers
- c_{i0} is the lower bound for the ith feature for the classifier c
- c_{i1} is the upper bound for the ith feature for the classifier c
- F_{xy} is the set of feature values for the pixel at (x, y)
- f_i is the value for the ith feature

3.5.1.2 Feature Extraction
A set of nine features are extracted for each pixel for use in classification. The features used are as follows:

1. Pixel greyscale value $f_1 = I(x, y)$.
2. Edge gradient magnitude (from Sobel), f_2.
3. Average of greyscale values in a 7×7 neighbourhood,

$$f_3 = \frac{1}{49} \sum_{\substack{x-3 \leq i \leq x+3 \\ y-3 \leq j \leq y+3}} I(i, j).$$

4. Greyscale value minus neighbourhood average,

$$f_4 = I(x, y) - \frac{1}{49} \sum_{\substack{x-3 \le i \le x+3 \\ y-3 \le j \le y+3}} I(i, j).$$

5. Greyscale value minus leaf blade average,

$$f_5 = I(x, y) - \frac{1}{|blade|} \sum_{\substack{0 \le i < width \\ 0 \le j < height \\ (i,j) \in blade}} I(i, j).$$

where *blade* is the set of all pixels which are part of the leaf's blade, found by using Otsu's thresholding (Otsu 1979) to remove the leaf from its background.

The average local gradient direction of pixels in an 11×11 neighbourhood around the current pixel is calculated. This size neighbourhood was chosen because for most vein pixels it will include both sides of the vein. The greyscale values of points five pixels from the current one in both directions along the gradient and perpendicular to the gradient are calculated. If the current pixel is part of a vein, the pixels perpendicular to the gradient direction are likely to also be vein pixels, and so similar to the current pixel, whilst the pixels along the gradient direction are likely to be non-vein, and therefore quite different.

$$i_1 = I(x + 5sin(\alpha), y + 5cos(\alpha))$$
$$i_2 = I(x - 5sin(\alpha), y - 5cos(\alpha))$$
$$j_1 = I\left(x + 5sin\left(\alpha + \frac{\pi}{2}\right), y + 5cos\left(\alpha + \frac{\pi}{2}\right)\right)$$
$$j_2 = I\left(x - 5sin\left(\alpha + \frac{\pi}{2}\right), y - 5cos\left(\alpha + \frac{\pi}{2}\right)\right)$$

where α is the gradient direction.

The remaining features are then:

6. The absolute difference between pixels, i_1 and i_2, either side of the potential vein
$$f_6 = |i_1 - i_2|$$

7. The absolute difference between pixels j_1 and j_2, along the potential vein
$$f_7 = |j_1 - j_2|$$

8. Greyscale value minus average value of the two pixels either side of the potential vein
$$f_8 = I(x, y) - \frac{i_1 + i_2}{2}$$

9. Greyscale value minus average value of the two pixels along the potential vein
$$f_9 = I(x, y) - \frac{j_1 + j_2}{2}$$

To allow the same genetic operators to be used on features with very varied distributions, the feature values for the training data are mapped to a uniform distribution. This mapping is recorded and applied to any data being subsequently classified.

3.5.1.3   Evolving the Classifiers

Classifiers are evolved one after another using a genetic algorithm and added to the classifier set until no more classifiers with a fitness above a certain threshold can be generated within a maximum number of iterations. The only genetic operators used are mutations, as crossover operations are likely to combine classifiers that work on different types of vein pixels, thereby having a negative effect. For example, a classifier that finds thin sections of vein may require higher edge gradient values and lower greyscale values than a classifier finding the pixels in the middle of thicker veins. Crossing over these two classifiers would result in one which classified neither of these vein pixel types. Bounds are mutated with probability 0.3 by adding or subtracting an amount randomly drawn from the range [0, 0.01]. The population is re-initialised after each classifier is added to the set. Each individual is initialised by centering the bounds around the feature values for a vein pixel randomly selected from the training data, with the width of the bounds drawn from a Gaussian distribution. This increases the likelihood of the classifier being effective, as one vein pixel will always be correctly classified by it, along with any similar vein pixels.

The fitness function used is as follows:

$$ fitness_i = \frac{|T_i \setminus \bigcup_{j \in C} T_j|}{|F_i \setminus \bigcup_{j \in C} F_j| + k}, $$

where:

- T_i is the set of vein pixels correctly classified by classifier i (true positives).
- F_i is the set of non-vein pixels incorrectly classified by classifier i (false positives).
- C is the current set of classifiers selected in previous iterations, and k is a constant.

This function grants high fitness to individuals which if added to the classifier set would significantly increase the number of true positives, but not the number of false positives. The fitness of a classifier is therefore dependent on the order in which they are selected. The constant k is used to adjust the balance between a high true positive/false positive ratio and a high total number of true positives. If k is set too low the ratio will be very high, but the final classifier set may over-fit the training data. If k is set too high it will result in a high number of false positives. A value of $k = 5$ was found to be appropriate.

3.5.1.4   Redundancy

Classifiers can potentially be made redundant by other classifiers added to the set later. In other words, a classifier may no longer uniquely classify many vein pixels whilst still incorrectly classifying some non-vein pixels. The removal of such classifiers is beneficial because the number of false positives may thus be greatly reduced whilst the number of true positives only slightly so.

Redundant classifiers are identified by removing candidates from the set and measuring any improvement in overall classification quality. The classifier whose

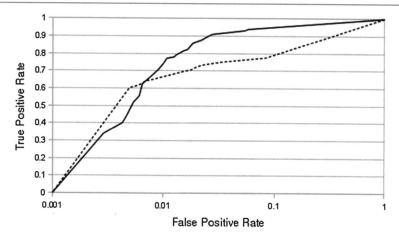

Fig. 3.11 ROC Curve. *Solid line* evolved classifiers. *Dashed line* ant algorithm

removal produces the largest increase in quality is permanently removed from the set. This process is repeated until no more classifiers are found to be redundant.

3.5.1.5 Results

The classifier was trained using 8000 pixels manually selected from 14 leaf images, two from each of seven species. These pixels were then manually labelled as either vein or non-vein. The resulting classifier was then tested on seven new leaf images, one from each of the species used for training. The ROC curve in Fig. 3.11 shows the results (solid line). With a false positive rate of 0.0166, a true positive rate of 0.853 was achieved.

The classifier was also used on the full leaf images from the test set, in order to extract the full venation pattern. Examples of these results are shown in Fig. 3.12.

3.5.2 Extraction by Ant Colonies

The second approach to vein extraction is to use an ant colony algorithm. A population of ant-like agents are placed at random across the image. These "ants" then move across the image from pixel to pixel based upon some heuristic evaluation of that pixel known as the pixel's visibility, and also based on the level of "pheromone" at that pixel. The pheromones are an indicator deposited by ants to signal to other ants the value of a particular pixel. As time progresses, the pheromone levels build up to create a pheromone map for the image, with high levels in desirable regions, and low levels in undesirable regions. In this case, the edge magnitude is used as the measure of visibility, to encourage the ants to traverse along the veins, thereby extracting continuous sections of venation.

Fig. 3.12 Results for
venation extraction by
evolved classifiers

(a) *Quercus shumardii*

(b) *Quercus rubra*

(c) *Quercus ellipsoidalis*

The probability, P_{ij}, of an ant at pixel i moving to pixel j is calculated as follows:

$$P_{ij} = \begin{cases} \dfrac{\tau_j^\alpha \eta_j^\beta}{\displaystyle\sum_{k \in K_i} \tau_k^\alpha \eta_k^\beta} & \text{if } j \in K_i \\ 0 & \text{otherwise} \end{cases}$$

where τ_j and η_j are the pheromone level and visibility respectively at pixel j, α and β are the weightings for these two components, and K_i is the set of pixels neighbouring pixel i.

To prevent the ants converging on the strong edges outlining the leaf instead of the venation, the visibility for all background pixels (again calculated using Otsu's method) and all pixels within a short distance of the background (in this case, a distance of 10 pixels) is set to 0. After all the ants have performed one move, the pheromone levels are updated:

$$\tau_{i+1} = (1 - \rho)\tau_i + \delta a_i \eta_i$$

where ρ is the rate at which pheromones evaporate, δ is the update rate, and a_i is the number of ants at pixel i.

There is a risk that ants will simply move between the same small set of pixels, building up pheromone levels until their escape becomes highly unlikely. This is prevented by keeping a list of the last 10 pixels visited by each ant and forbidding the ant from re-visiting any of these pixels. After a set number of moves have taken place the pheromone map is thresholded to produce a binary vein classification.

3.5.3 Results and Comparison of Methods

Figure 3.13 contains examples of typical results obtained using this method. For each leaf the algorithm was run for 500 steps, using 2000 ants. The pheromone map was then thresholded at 2% of the maximum pheromone level. These values were chosen as they appeared to give the best qualitative results. The results differ from those obtained using the evolved classifiers in a number of ways. Firstly, due to the use of only the edge gradients to guide the ants across the image, the results contained only the hollow outline of the venation, whereas the other method extracted the full vein. One advantage of using ants is that it helps in extracting continuous venation, whilst the evolved classifiers extract veins with many small gaps in them. On the downside, when a vein contains a section with only a low edge magnitude, the ants are unable to continue to extract the rest of that vein as the pixel-by-pixel evolved classifiers are able to do. The effect of this can be seen near the top of the first image in Fig. 3.13, where a large section of venation is completely absent. Furthermore, whereas many of the false positives resulting from the first method were isolated pixels that can easily be removed, the ants produced larger, connected areas of noise, which are harder to distinguish from the actual venation.

By applying morphological closing, the hollow vein centres can be filled in (Fig. 3.14) and quantitative results can then be calculated, as shown in Fig. 3.11

Fig. 3.13 Results of venation extraction using the ant colony algorithm (same leaves as Fig. 3.12)

Fig. 3.14 Results of venation extraction after morphological closing (same leaves as above)

(dashed line). It can be seen that the ant algorithm still performs more poorly than the evolved classifiers, except when the true positive rate falls below approximately 0.63.

3.6 Summary

This chapter describes techniques for the extraction of some of the key components of plant leaf morphology (primarily venation, margins and texture), providing appropriate descriptors which can be used in the automated comparison and classification of leaves into their species. In addition, the results of a comparative study of the most popular techniques for analysing leaf shape are presented. These show that both elliptic Fourier descriptors and shape features perform well for leaf classification, but while the former work better for simpler outline shapes, the latter are preferable for leaves which differ more strongly in shape and have more complex outlines.

Machine Learning for Plant Leaf Analysis

4

Chapter contributions:

- Incorporating within-species variance into classification methods for leaf blades and margins
- Probabilistic classification combining multiple leaf feature-sets
- A technique for automatic selection among feature-sets on a leaf-by-leaf basis.

4.1 Incorporating Intra-species Variation into Plant Classification

One of the key challenges for automated analysis of plant leaves lies in the range of variation presented by a species and even by a single plant (Fig. 2.1). As well as the natural variation to be expected from any organic object, the variation of a leaf can arise from a number of sources, for example, its age and developmental stage. Shape varies during development, with early growth phases occurring primarily length-wise and increase in width coming later. In some species with lobed leaves, the leaf lobes are not apparent until after a certain stage in development. In others, like many *Eucalyptus* taxa, the leaves of young shoots are morphologically very distinct from those of mature ones. Margin characteristics such as teeth may not develop until the leaf has reached full size, often appearing first near the apex and then gradually extending further back towards the insertion point. Pigmentation often changes as the leaf develops.

Another source of variation arises from damage, usually resulting from disease or insect attack. Diseases commonly affect the surface of the leaf, ranging from discoloration to distinct marking, while insects often alter a leaf's shape by eating parts of it.

© Springer-Verlag GmbH Germany 2017
P. Remagnino et al., *Computational Botany*, DOI 10.1007/978-3-662-53745-9_4

Much variation can also be introduced by the image capture process. Lighting conditions may play a large role here. Many leaves have waxy surfaces which may reflect light differently, depending on the relative position of the lighting source, and the amount of light transmitted through the leaf usually affects the visibility of features such as the fine venation. Some leaf data sets have been created using a specific backlighting system in order to emphasize this feature. Camera focus and resolution also affects the amount of texture information available for capture. Cameras are not the only devices that have been used to capture leaf images, other examples including flat-bed scanners, x-ray devices and even electron microscopes.

This section explores methods for increasing the reliability of leaf classification by taking account of intra-species variation that may be present in the dataset.

4.1.1   Utilizing the Hungarian Algorithm for Improved Classification of Leaf Blades

For the leaf macro-texture, data of the type generated in Sect. 3.2.1 - where the leaf to be classified is described by a distribution of points within a feature space - can be classified using a number of existing methods.

When described using histograms, the difference between two probability density functions (pdfs) can be calculated using bin-by-bin methods such as the Jeffrey-divergence metric. However, these methods encounter problems if the data has a high dimensionality, because a large number of bins makes the calculation expensive, but a sparse population of bins produces poor results. The earth mover's distance (EMD) (Rubner et al. 1997) deals with this by using signatures, and provides an accurate and intuitive measurement. These 'signatures' are weighted points within the feature space. This is akin to clustering data points drawn from the distribution and weighting each cluster centroid by the number of points in the cluster. Another method is to use kernel density estimation (Parzen 1962) to estimate a probability density function using points sampled from a distribution, and then to use this estimation to predict the probability of another sampling of points belonging to the same distribution. More recently, 'bag-of-words' methods have enjoyed increasing usage for this problem, particularly in the guise of 'bag-of-visual-words' (Sivic and Zisserman 2003) for image retrieval.

To overcome the problems inherent in leaf macro-texture variation, two different methods are presented here which utilize information generated in calculating the earth mover's distance in order to allow for more robust classification of pdfs, particularly when there is high intra-class variation. The first of these methods combines this with the strengths of the 'bag-of-words' method, while the second uses this information more directly to try to model the intra-class variation.

4.1.1.1   Background
The Hungarian Algorithm and The Earth Mover's Distance

Fig. 4.1 The mapping
produced by the Hungarian
algorithm between two sets
of points

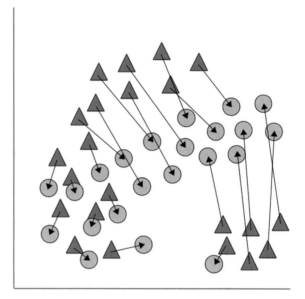

The earth mover's distance (EMD) (Rubner et al. 1997) is a measure of the difference
between two pdfs. The analogy is that to reform one mound of earth as another, the
effort required would depend on the sum of the distances that each unit of dirt
must be moved. Whilst bin-by-bin methods only consider the amount of 'earth' in
each location, the EMD considers how far it must be moved. There are two forms
of pdf descriptions that allow the EMD to be calculated: histogram binning and
the aforementioned signatures. Since binning is analogous to using evenly spaced
signatures, only the latter needs to be considered.

While there may be many ways of reforming one pdf into another, the EMD
requires the minimum total movement (the sum of the distances that each unit of
'earth' is moved). See Fig. 4.1 for an example of how one set of data can be mapped
to another in this manner. The standard way of determining this is to model it as the
transportation problem – the assignment of sources to destinations subject to a set of
transportation costs. There are a number of methods for solving the transportation
problem, but by reforming the data so that each signature has an equal weight, it
becomes equivalent to the simpler assignment problem, which can be solved using
the Hungarian algorithm (Kuhn 1955). Whilst the original Hungarian algorithm was
$O(n^4)$, an $O(n^3)$ version has since been found by Edmonds and Karp (1972).

The EMD uses only the minimum cost calculated by the Hungarian algorithm, but
here the corresponding mapping between signatures is also recorded, as it provides
not only a measurement of the difference between the pdfs, but also information
about their differences. The EMD normally uses the Euclidean distance as the cost
of moving 'earth' between two points, but here the squared Euclidean distance is
used, as this helps to preserve the topology/ordering of the points (Fig. 4.2), since
the pairing of points over increasing distances is penalised.

Fig. 4.2 Using the squared
Euclidean distance as the
cost function preserves the
topology

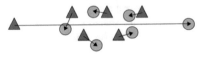

(a) Euclidean distance

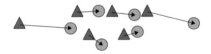

(b) Squared Euclidean distance

The Bag-of-Words Model

The 'bag-of-words' model was originally used for the retrieval of text documents (Sparck-Jones and Needham 1968). The idea was to represent documents as the frequency of occurrence of different words and to discover similar documents by comparing these frequencies. In recent years this concept has been extended to the classification of more general forms of data. Typically, a large number of points are sampled from the training distributions and subjected to clustering. The cluster centroids are then treated as 'codewords' in a 'dictionary' used to perform a quantization of the data by assigning each data-point to its nearest 'codeword'. A set of points from a new distribution can then be described as the frequency of occurrence of each 'codeword'. This concept has seen much recent use in the field of computer vision for tasks such as image retrieval (Sivic and Zisserman 2003; Tirilly et al. 2008; Chen et al. 2009) and texture analysis (Leung and Malik 2001; Varma and Zisserman 2009).

4.1.1.2   Notation

The problem is defined as follows. A leaf is described by a set of n data points, $X = \{\bar{x}_1, \bar{x}_2, \ldots \bar{x}_n\}$, consisting of windows sampled from the leaf's surface. Each data point $\bar{x}$ is a feature vector, $\bar{x} = [x_1, x_2, \ldots, x_d]$, where d is the number of features. Given a number of different species, where species i is described by another set of n data points, $C_i = \{\bar{y}_1, \bar{y}_2, \ldots \bar{y}_n\}$, randomly sampled from all leaves in the training set that belong to the species, the aim is to determine the species to which the leaf described by X most likely belongs. This is calculated using Bayes theorem:

$$c^* = \arg\max_{C_i} P(C_i|X) \tag{4.1}$$

$$= \arg\max_{C_i} P(X|C_i)P(C_i)/P(X) \tag{4.2}$$

$$= \arg\max_{C_i} P(X|C_i)P(C_i) \tag{4.3}$$

The term $P(X)$ in Eq. 4.2 is discarded as it is constant for all i.

4.1.1.3 Data-Point-Mapped Bag-of-Words

The method involves first generating a set of codewords from the training set, suitable for representing the data. All points in the training leaf and species objects are assigned to their nearest codeword. A mapping is calculated between the data points in each training leaf object and its corresponding species object. For each species, the joint distribution is calculated for a training object point assigned to one particular codeword being mapped to a species object point that is assigned to a second codeword. That is, for each pair of codewords and each species, the probability is calculated of a mapping having its training object point assigned to the first of these codewords and its species point assigned to the second. For classification, the same codeword assignments and mappings are performed, and the previously calculated probabilities are used to determine the species to which the leaf belongs.

Generating a Dictionary

Within the literature there has been much discussion on the best size and appropriate methods for generating the codeword dictionary (Jurie and Triggs 2005; Chatfield et al. 2011; Sikka et al. 2012). The simplest approach is to choose evenly distributed points throughout the feature space. This has the disadvantage that large portions of the space may not be used, resulting in redundant codewords, while other more useful areas may be inadequately represented. A simple method that largely eradicates this problem is to use randomly selected points from the training data as the codewords. However, using the centroids from a clustering performed on the training data normally provides a better representation. Another approach is to perform a separate clustering for each class and then combine the generated codewords. This ensures that each class has some appropriate codewords, but may result in very similar codewords in the combined dictionary. It was found that a k-means clustering of the whole training set produces an appropriate dictionary for this method.

There is no consensus on the size of a dictionary, but for the method used here it was found that for objects described using 1024 points, a dictionary of size 256 produced good results. Larger dictionaries provided little or no improvement. D_i is the ith codeword in the dictionary.

Producing the Class Models

For each species, a species object is produced by randomly selecting n points from the examples of that species in the training set. For each training leaf, a mapping is found from its data points to those of its species object using the Hungarian algorithm. This mapping pairs the points in one object to those in the other, such that the sum of the squared Euclidean distances between paired points is minimised (see Fig. 4.1). The point in the species object C_i to which point $\bar{x}$ is paired is defined as $M(\bar{x}, C_i)$.

Each point in the training data is assigned to its nearest codeword. For each species i and for each pair of codewords, (D_a, D_b), the conditional probability is calculated of a point $\bar{x}$ in the training data of that species being assigned to codeword D_a, given that the corresponding point in the species object has been assigned to D_b. This is calculated as follows:

$$P(W(\bar{x}) = D_a | W(M(\bar{x}, C_i)) = D_b) \qquad (4.4)$$

$$= \frac{P(W(\bar{x}) = D_a, W(M(\bar{x}, C_i)) = D_b)}{P(W(M(\bar{x}, C_i)) = D_b)} \qquad (4.5)$$

where

$$P(W(\bar{x}) = D_a, W(M(\bar{x}, C_i)) = D_b) = \sum_{\substack{W(T_{ij})=D_a \\ W(M(T_{ij},C_i))=D_b}} \frac{1}{|T_i|} \qquad (4.6)$$

$$P(W(M(\bar{x}, C_i)) = D_b) = \sum_{d=0}^{|D|} P(W(\bar{x}) = D_d, W(M(\bar{x}, C_i)) = D_b) \qquad (4.7)$$

where T_{ij} is the jth point and $|T_i|$ is the total number of points in the training data for species i, $|D|$ is the number of codewords, and $W(\bar{x}) = D_a$ indicates that point $\bar{x}$ has been assigned to codeword D_a (likewise, $W(M(\bar{x}, C_i)) = D_b$ indicates that the point to which $\bar{x}$ is paired is assigned to codeword D_b).

Equation (4.6) calculates the probability of a point in D_a being mapped to a point in D_b as the fraction of training points for a species C_i for which this occurs. The probability of a point, from any codeword, being mapped to one in D_b is then the sum of these for all codewords (Eq. (4.7)).

Performing The Classification

To classify a leaf, all the data points of the leaf are assigned to determine their nearest codewords. The object is mapped to each of the species objects using the Hungarian algorithm. The species to which the leaf most likely belongs can then be determined using a Bayesian classifier.

$$c^* = \underset{C_i}{\arg \max}\, P(X|C_i)P(C_i) \qquad (4.8)$$

$$P(C_i) = \frac{|T_i|}{\sum_j |T_j|} \qquad (4.9)$$

$$P(X|C_i) = \prod_{W(\bar{x})=X} P(W(\bar{x}) = D_a | W(M(\bar{x}, C_i)) = D_b) \qquad (4.10)$$

Fig. 4.3 Descriptors are generated to model the movement between the class object and another object in terms of each cluster

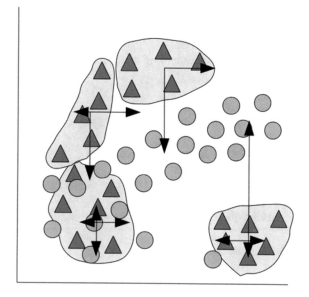

4.1.1.4 Intra-class Variation Models

The second method attempts to improve reliability by modelling the variation within each species. The data points of each species object are separated into a number of clusters. We model the movement (in the transformation from one pdf to another) within each of these clusters under the mapping between the species object and its training examples (Fig. 4.3). This essentially aims to describe how each portion of the distribution typically varies for that species. These models are then used to determine to which species another leaf most likely belongs.

Training the Classifier

For each species, a species object is created as before. Next a small number, k, of cluster centroids are found for each species, using the clustering algorithm described in Fig. 4.4. This method of clustering creates clusters of equal size, and thereby helps to ensure the centroids are appropriately spread throughout the distribution, with centroid density approximately proportional to the density of the data points. Any clustering algorithm with similar properties could also be used. All the points in the species object are then assigned to the cluster of their nearest centroid. The change between objects will be measured relative to these clusters. The jth cluster for species i is denoted as K_{ij}.

For each training leaf object X_i^t for species i, a movement descriptor F_i^t is generated, after the species object C_i has been mapped to the training object (as before, using the Hungarian algorithm). Each element in the descriptor is the sum of the movements of points within a particular cluster for a particular dimension and in a particular direction.

Initialise cluster centroids at randomly picked data points
repeat
   for all clusters **do**
      Sort points according to distance from centroid
   end for
   repeat
      for all clusters **do**
         Assign next nearest unassigned point to cluster
      end for
   until all points assigned to clusters
   for all clusters **do**
      Calculate centroid as mean of points in cluster
   end for
until sufficiently converged, or max iterations reached

Fig. 4.4 Clustering algorithm

$$F_i^t = \{\bar{f}_{ab}^t | 0 < a < k; 0 < b < d\} \tag{4.11}$$

$$\bar{f}_{ab}^t = [f_{ab+}^t, f_{ab-}^t] \tag{4.12}$$

$$f_{ab+}^t = \frac{k}{|X_i^t|} \sum_{\substack{\bar{x} \in X_i^t \\ x_b - M(\bar{x}, C_i)_b > 0 \\ M(\bar{x}, C_i) \in K_{ia}}} x_b - M(\bar{x}, C_i)_b \tag{4.13}$$

$$f_{ab-}^t = \frac{k}{|X_i^t|} \sum_{\substack{\bar{x} \in X_i^t \\ x_b - M(\bar{x}, C_i)_b < 0 \\ M(\bar{x}, C_i) \in K_{ia}}} x_b - M(\bar{x}, C_i)_b \tag{4.14}$$

where d is the number of dimensions, k is the number of clusters and t is the training instance. $\bar{x} \in K_{ia}$ indicates that point $\bar{x}$ is assigned to the ath cluster for species i, and x_b refers to the value in $\bar{x}$ corresponding to the bth feature (likewise for $M(\bar{x}, C_i)_b$).

Equations (4.13) and (4.14) calculate the elements of the descriptor for cluster a, in the positive ($x_b - M(\bar{x}, C_i)_b > 0$) and negative ($x_b - M(\bar{x}, C_i)_b < 0$) directions, respectively, along dimension b. These are calculated as being the sum of the distances between training points and their mapped species points, where the mapped point is in the given cluster ($M(\bar{x}, C_i) \in K_{ia}$). These are normalized by multiplying by the number of clusters, k, divided by the number of points in the training object ($|X_i^t|$).

Classification

To classify a leaf X for each potential species, the mapping and generation of a movement descriptor, F_i, is performed as in the training stage. We then use a Parzen window method (Parzen 1962) with a Gaussian kernel to calculate the likelihoods for each species and determine the classification.

$$c^* = \arg\max_i P(X|C_i)P(C_i) \tag{4.15}$$

$$P(X|C_i) = P(F_i|C_i) \tag{4.16}$$

$$= \prod_{a=0}^{a<d} \prod_{b=0}^{b<k} P(\bar{f}_{ab}|C_i) \tag{4.17}$$

$$P(\bar{f}_{ab}|C_i) = \frac{1}{T_i} \sum_{t=0}^{T_i} P(\bar{f}_{ab}|\bar{f}_{ab}^t) \tag{4.18}$$

$$P(\bar{f}_{ab}|\bar{f}_{ab}^t) = \phi(||\bar{f}_{ab}^t - \bar{f}_{ab}||) \tag{4.19}$$

where T_i is the number of training examples for species i and $\phi(x)$ is a normal distribution function with mean, $\mu = 0$ and standard deviation, $\sigma = 0.002$.

4.1.1.5   Experiments

In this section the new algorithms are empirically evaluated by comparing them to a selection of other techniques. For these experiment we have 32 different species, with 16 examples of each, performing a 16-fold cross validation. The object of each example has 1024 data points, generated as described in Sect. 3.2.1. For the first method (Sect. 4.1.1.3) we used dictionaries of up to 256 codewords, and for the second (Sect. 4.1.1.4) we used up to 64 clusters for each species. Whilst better results were obtained with increasing numbers of codewords and clusters up to these limits, no further significant improvement was found above them.

Methods for Comparison

The three methods we used for comparison were kernel density estimation, the earth mover's distance, and a bag-of-words method using a Naive-Bayes classifier.

- Kernel Density Estimation:
  Kernel density estimation is used to predict the probability density function for each species. This estimate of the pdf is then used to calculate the likelihood that a given leaf belongs to that species.

$$P(X|C_i) = \prod_{\bar{x} \in X} P(\bar{x}|C_i)$$

$$= \prod_{\bar{x} \in X} \sum_{\bar{y} \in C_i} \frac{\phi(||\bar{y} - \bar{x}||)}{|C_i|}$$

where $\phi(x)$ is a normal distribution function with mean, $\mu = 0$ and standard deviation, $\sigma = 0.1$. This kernel function was used as it appeared to give the best results for the dataset.

Table 4.1 Results for each method (% correctly classified)

(a) Method 1, data-point-mapped bag-of-words, varying object and dictionary size (in %)

	$n = 256$	$n = 512$	$n = 1024$
$|D| = 16$	67.97	73.05	75.39
$|D| = 32$	75.39	80.66	81.64
$|D| = 64$	84.77	85.35	88.09
$|D| = 128$	86.13	90.04	90.06
$|D| = 256$	90.02	91.02	92.97

(b) Bag-of-words method, varying object and dictionary size (in %)

	$n = 256$	$n = 512$	$n = 1024$
$|D| = 16$	57.03	63.28	63.28
$|D| = 32$	62.70	65.82	67.19
$|D| = 64$	69.73	74.02	74.02
$|D| = 128$	74.41	76.76	77.54
$|D| = 256$	77.15	79.30	80.27

(c) Method 2, intra-class variation models, varying object size and number of clusters (in %)

	$n = 256$	$n = 512$	$n = 1024$
$k = 8$	69.73	80.08	86.33
$k = 16$	83.79	90.63	92.97
$k = 32$	91.21	93.75	98.05
$k = 64$	94.73	96.48	98.05

- Earth Mover's Distance:
  For this we used the pure value calculated by the earth mover's distance instead of utilizing the mapping between objects. Each leaf is classified as belonging to the species whose object is closest to it according to the EMD metric.
- Naive-Bayesian Bag-of-Words:
  For the bag-of-words method, we used the same codeword dictionary as for the new method (Sect. 4.1.1.3), to allow fairer comparison. We used a Naive-Bayes classifier, as it is both one of the most common classifiers (along with SVMs) used for bag-of-words (Csurka et al. 2004), and is similar to that used in the proposed method.

Results

Table 4.1a, b, c give the results, respectively, for the first proposed method (Sect. 4.1.1.3), bag-of-words method, and second proposed method (Sect. 4.1.1.4), using different numbers of data points and different dictionary sizes/numbers of clusters. Exactly the same dictionaries were used for the first method and bag-of-words method. The overall results of the experiments are given in Table 4.2.

Table 4.2 Overall results, using best parameter values for each method (% correctly classified)

Method	$n = 256$	$n = 512$	$n = 1024$
First proposed method	90.02	91.02	92.97
Second proposed method	94.73	96.48	98.05
Kernel density estimation	69.73	73.83	77.73
Earth mover's distance	73.83	79.88	85.35
Bag-of-words	77.15	79.30	80.27

As the results show, the new methods both performed far better than the standard bag-of-words method. This is because when the difference between pdfs means that points are assigned to different codewords, the standard method considers only that these points are no longer assigned to the same codeword, whereas the new methods both consider where in the feature space those points may be located, given a particular class. The kernel density estimation and earth mover's distance methods both performed worse than the other methods because they directly compare samplings from distributions and so are susceptible to noise produced by the sampling. The bag-of-words methods eliminate much of this noise by quantisation via assignment to codewords, as does the second new method, but instead by using the behaviour of different parts of the distribution.

Of the two new methods, the second performs better for plant leaf classification, with 98.05% of leaves correctly classified versus 92.97%. This is probably because it deals better with within-species variation which for this dataset may be quite high due to varying levels of damage or disease present on the leaves and to slight differences in lighting conditions. For other data where there is either less intra-class variation or it is less quantifiable, it is possible that the first method may perform better.

Given that the EMD must be calculated to carry out the new methods, it may be possible to improve the results by incorporating the EMD metric. In this case, however, doing so produced no change in results. As expected, increasing the number of points used to describe objects increases the quality of the classification, but the new methods still perform better than the others when a smaller number of points are used, making them particularly suitable when larger samplings are not practical.

Due to the $O(n^3)$ nature of the Hungarian algorithm, the method presented here can be quite slow compared to some others, requiring approximately 6 seconds per leaf in these tests, with $n = 1024$. Despite this, for many applications the additional time required is entirely acceptable given the improvement in accuracy, and the cost can be mitigated to some extent, for example, by using a faster, less reliable method to eliminate the least likely classes first. Furthermore, these methods still performed better than the others tested here when $n = 256$, greatly reducing the time required and providing accurate classification when less data is available.

4.1.2 Comparing Leaf Margins Using Dynamic Time Warping

Within some species much variation also occurs in the leaf margin, largely due to differences in teeth size and their extent of occurrence along the margin, as the leaf develops. To compare two margins, a common starting point must first be selected and the obvious candidates are the apex and insertion point (Fig. 1.1).

Although these two points have been identified (in Sect. 3.4), the procedure does not identify which is which. Consequently, when performing a comparison, all four combinations (possible pairings) for sequence start points are used. In the case of asymmetric leaves, the distance between insertion point and apex may be quite different even though the details of the margins are similar on each side. To take account of this, the margin signatures (generated as described in Sect. 3.3) were oriented always to proceed along the shortest side first. The DTW algorithm (Sect. 3.4.1) was applied for all four configurations and the smallest measurement was selected as being the difference between the two leaf margins.

Following the assumption in Sect. 3.4 that the maximum length difference between the two sides of the leaf will be $2w$, where $w = \frac{n}{8}$, the value of k used in Eq. (3.6) was also set to $\frac{n}{8}$, since this was the point in the sequences where most distortion was expected to occur.

4.1.2.1 Results

The method was evaluated on a dataset containing sixteen leaves from each of 100 different species. A 16-fold cross-validation was performed, such that one leaf from each species was used each time in the testing set, whilst the remaining leaves comprised the training set. Classification was performed using the k-nearest-neighbour technique, with $k = 5$; this value was chosen because it was found to produce the best results. For comparison, two other techniques were also used on the same data:

- Cross-correlation:
  For two sequences, $\mathbf{a} = \langle a_1, \ldots, a_m \rangle$, $\mathbf{b} = \langle b_1, \ldots, b_m \rangle$, the distance between the two was calculated for every possible offset of one sequence against the other. The shortest distance calculated was used for the classification.

$$distance = \min_{0 \le i < n} \sum_{j=0}^{n} \| a_j - b_{j+i \bmod n} \|$$

- Bag-of-Words:
  A large number of feature vectors were sampled from the entire training set and a k-means clustering performed on them. The cluster centroids constituted the codewords of a dictionary used to perform a quantization of the data, by assigning each data-point to its nearest codeword. A margin sequence can then be described as the frequency of occurrence of each 'codeword' (Csurka et al. 2004). For classification, the distance between two sequences was then calculated as the Jeffrey-divergence metric for their two histograms.

Table 4.3 Results for the three methods of leaf margin classification

Method	Result (% correctly classified)
Cross-correlation	57.12
Bag-of-words	74.51
Proposed method	91.32

As the results show (Table 4.3), the proposed method performed significantly better than the other two. The cross-correlation method conserves the order of the sequence but is too rigid to account for the variation that occurs in leaf margins, for example in the exact positions of the tooth tips which appear as peaks in the signature. By ignoring the order of the sequence, the bag-of-words method describes only the content of the margin, and loses valuable information. By using the DTW algorithm, the proposed method can utilize the order of the sequence, whilst having enough flexibility to deal with the variation inherent to natural data.

4.2 Combining Different Leaf Features

Although it has been shown previously that it is possible to achieve high levels of accuracy in classification using single features of plant leaves, it seems clear that further improvements could be obtained by combining multiple leaf components. With greater dataset size and consequent increase in overall variation, some features will no longer be effective in distinguishing the species, due to increased overlap between groups within the search space brought about by greater relative within-species/among-species variation. For example, the leaves of many different species have a similar oval-shaped blade (Fig. 4.5) and alone this feature may be insufficient for species discrimination, but combining it with texture information may prove sufficient for accurate classification.

In addition to obtaining a robust classification method by combining different leaf feature-sets, it is useful to assess the utility of each feature-set, not only for accuracy in general classification but also to determine its usefulness for classifying a particular leaf. Given the computational costs of extraction and comparison, an assessment is desirable as to whether all of the leaf components and methods available are needed. It is also possible that some feature-sets are strongly correlated with each other, thus reducing the value of using both together. Another consideration is the choice between using a small selection of highly rigorous but computationally expensive methods, or a larger number of simpler but faster methods.

Fig. 4.5 Many species have similar leaf shapes

4.2.1 Probabilistic Classification from K-Nearest-Neighbour

Here, the use of k-NN classifiers for producing probabilistic classifications from multiple leaf feature-sets is explored. Typically with k-NN classifiers, a test object is assigned to the class which prevails among its k nearest neighbours from the training set. However, a number of methods have been proposed for producing a probability for each class from the set of neighbours. The two methods investigated here are those of Fukunaga and Hostetler (1975) and an extended version of Atiya (2005).

4.2.1.1 Posterior Probability Estimation

For each species/class, c_i, the probability of a leaf, x, belonging to it is calculated as

$$P(c_i|x) = \frac{\prod_{f=1}^{F} P_f(c_i|x)}{\sum_{i=0}^{C} \prod_{f=1}^{F} P_f(c_i|x)}$$

where $P_f(c_i|x)$ is the probability of class c_i for leaf feature-set f, and F is the total number of feature-sets used.

Fukunaga's method

This method calculates the probability from feature-set f as:

$$P_f(c_i|x) = \frac{K_i}{K}$$

where K is the total number of neighbours being used, and K_i is the number of neighbours belonging to class c_i.

Atiya's method

Atiya extended Fukunaga's method to include weights, calculated from the training set.

$$P_f(c_i|x) = \sum_{j=0}^{K} v_j B_{ij}$$

where v_j is the weight for the jth neighbour, and B is a matrix with $K + 1$ columns and C rows, with B_{ij} set to 1 if the jth neighbour is from class i and 0 otherwise. The elements in the final column are all set to $\frac{1}{C}$.

The weights $v_j, j = 1..K$ are calculated as

$$v_j = \frac{e^{w_j}}{\sum_{k=1}^{K} K e^{w_j}}$$

with w_j determined by maximising the likelihood of the data. Each value is initialised to be equal and is then updated by

$$w_j = w_j + \eta \left[\sum_{n=1}^{N} \frac{B_{c_n j}(n) e^{w_j}}{\sum_{i=1}^{K+1} B_{c_n i}(n) e^{w_i}} - \frac{N e^{w_j}}{\sum_{i=1}^{K+1} e^{w_i}} \right]$$

where $B(n)$ and c_n are the B matrix and class, respectively, for the nth training sample, and η is the step size. This update is repeated until the change in weights becomes negligible.

4.2.1.2   Experiments

To evaluate these two methods, four different leaf feature-sets were used:

1. Shape features (SF) - a set of eight features as described in Sect. 3.1.1.1.
2. Elliptic Fourier descriptors (EFD) - as described in Sect. 3.1.1.3. EFD were used as well because they can differentiate species pairs that shape features cannot distinguish.
3. Margin histogram - a 32-bin histogram was generated by quantizing the data generated in Sect. 3.3.
4. Blade histogram - the histograms generated in Sect. 3.2.1 were used.

Due to the superior results obtained when using the blade and margin as compared to the shape, a smaller dictionary and sampling sizes were used for the data here, the better to show the advantage of combining multiple feature-sets.

A 16-fold cross-validation was performed on the 100-species dataset (16 samples per species). Results were generated for every combination of the four leaf feature-sets and are shown is Table 4.4. The accuracy value for a row represents the fraction of leaves correctly classified using the method and feature-set combination given in

Table 4.4 Results for each combination of method and leaf feature-sets

Method	Shape	EFD	Blade	Margin	Accuracy	Deviation
Fukunaga			✓		0.5456	0.2739
	✓				0.5987	0.3108
		✓			0.6056	0.2897
				✓	0.6825	0.2796
	✓	✓			0.7531	0.2609
		✓		✓	0.8625	0.1709
		✓	✓		0.8731	0.1359
	✓		✓		0.8756	0.1315
	✓			✓	0.8831	0.1626
	✓	✓		✓	0.8937	0.1579
	✓	✓	✓		0.9093	0.1119
			✓	✓	0.9143	0.1048
		✓	✓	✓	0.9587	0.0656
	✓	✓	✓	✓	0.9625	0.0712
	✓		✓	✓	0.9643	0.0601
Atiya			✓		0.6437	0.2344
		✓			0.6593	0.2577
	✓				0.6643	0.2779
				✓	0.7212	0.2345
	✓	✓			0.7887	0.2249
		✓		✓	0.8762	0.1533
	✓		✓		0.8925	0.1182
	✓			✓	0.8981	0.1437
	✓	✓		✓	0.9025	0.1458
		✓	✓		0.9050	0.1184
	✓	✓	✓		0.9262	0.1098
			✓	✓	0.9337	0.0717
		✓	✓	✓	0.9681	0.0602
	✓	✓	✓	✓	0.9681	0.0640
	✓		✓	✓	0.9688	0.0553

that row. The deviation is the standard deviation for the accuracy between different species. A high deviation shows that the feature-set combination of that row performs much better for some species than for others, while a low deviation shows that it works similarly well for all species.

As expected, in both cases accuracy generally improved with increase in the number of feature-sets used. Atiya's method slightly outperformed Fukunaga's, indicating the value of weighting the contribution of each neighbour. It is worth noting that when all the feature-sets except EFDs were used, accuracy was slightly higher than for all four feature-sets. This shows that sometimes the inclusion of certain features can be detrimental to the result.

Of the single feature-set cases, the margin feature-set performed best, while the blade features performed the worst. Despite this, the three feature-set combinations in which the blade features were used in combination with only one of the shape-based feature-sets (SF or EFD) achieved significantly better results than when combined with both shape-based feature-sets, illustrating the advantage of using a diverse set of features.

4.2.2   Automatic Feature Selection

As noted previously, leaf feature-sets differ in their ability to differentiate (and hence classify) species, according to the nature and degree of variation the species samples exhibit in the characters being used. When in a given species pair, within-species variation exceeds among-species variation in a feature-set, these features may not be able to distinguish the two species. Indeed, the use of such features could be detrimental to achieving the correct classification. For example, it is possible that for a given feature-set, none of the nearest neighbours of the test object belong to the correct species, resulting in an incorrect classification, irrespective of the values of other features. It therefore may not always be best to use all features available. This can be seen in Table 4.4 where some three feature-set combinations performed better than those with four feature-sets.

As datasets increase in size (it has been estimated that there may be up to 400,000 plant species) the computational cost of identifying a species could become very great. Every feature-set used would add to this cost and so there would be a clear benefit were it possible to reduce the number of features necessary for classification.

Much work has been done in the field of feature selection. Most techniques aim to find suitable subsets, either by searching through candidate subsets (Wrapper methods) or by using prior knowledge to predict the best features (Filter methods). Wrapper methods range from basic techniques such as forward selection and backward elimination (Kittler 1978) to more modern methods such as those of Chen et al. (2011) and Rashedi et al. (2013). Filter methods include correlation-based selection techniques (Yu and Liu 2004) and Markov blanket filters (Koller and Sahami 1996).

Because the success of leaf feature-sets in species classification is correlated with the sampled patterns of variation, it may be beneficial to dynamically select the best feature-sets for each leaf, rather than using a predetermined combination. Furthermore, it may prove possible to evaluate the utility of a particular component for classifying a particular leaf without needing to generate a full set of features for it. Here, a method and a number of metrics are explored for dynamically selecting feature-sets on a leaf-by-leaf basis, along with an evaluation of their effectiveness.

4.2.2.1 Metrics for Feature Utility

Given a vector for a particular leaf component, there are a number of different metrics which could be used to estimate that component's utility for classifying the leaf, prior to performing the classification. The metrics explored here are based on the neighbourhoods (the sets of nearest neighbours) used in the previous section for classifying the leaves. These metrics are calculated for all examples in the training set. When a new vector is presented, the methods to be described in Sect. 4.2.2.2 use these past calculations to estimate the value for the new vector.

1. *Same versus K* - The fraction of the K nearest neighbours which belong to the same species/class as the training example – K_t/K, where K_t is the number of neighbours from the same species as the training example. This metric reflects Fukunaga's method in Sect. 4.2.1.1.
2. *Same versus Next Highest* - The number of neighbours from the correct class divided by the number of neighbours for the second highest scoring class. This gives a measure of the likelihood that the vector would have been correctly classified.
3. *Neighbouring Classes* - Metric 1 divided by the total number of different classes represented within the neighbourhood. If all the neighbours are from different classes, the vector is less likely to be classified correctly than if they are all from the same class.
4. *Entropy* - Provides a measure of uncertainty, with low values indicating a high level of predictability. Calculated as

$$E = 1 - \frac{\sum_{i=1}^{C} p(c_i) \log p(c_i)}{\log K}$$

where $p(c_i)$ is the fraction of neighbours belonging to class c_i. This metric has the advantage of giving the same value for a vector in relation to its neighbours, regardless of the species to which it belongs.

4.2.2.2 Estimation of Feature Utility

For estimating the utility of a new vector, a feed-forward neural-network is trained via back-propagation on the training vectors with the utility metric values for those training vectors as the expected output. The network consists of two hidden layers, each with twice as many nodes as the number of input nodes and a single output node. The number of input nodes is dependent on the leaf feature-set being used. Sigmoid functions ($f(x) = (1 + e^{-x})^{-1}$) are used at each hidden and output node. Training was performed by introducing all of the training vectors to the network in a random order. This process was repeated until the decrease in the average output error dropped below a small threshold. Once trained to sufficient convergence, the utility of a new vector can be estimated by inputting the vector to the trained network.

4.2.2.3 Classification with Leaf Feature Selection

Once an estimate of the utility of each of the four feature-sets for a leaf has been obtained using the trained neural-network, it can be used to minimise the number of feature-sets required to perform a classification. The leaf is first classified using the feature-set with the highest utility and the probability for the top result is calculated as described in Sect. 4.2.1. If this probability is greater than some predetermined threshold (to be discussed in Sect. 4.2.2.4), the result is accepted, but if not the feature-set with the next highest utility is selected and the probability recalculated, until the threshold is passed.

In cases in which all the available features appear to be necessary, the contribution from each feature-set is weighted according to its estimated utility:

$$P(c_i|x) = \frac{\prod_{f=1}^{F} P_f(c_i|x)^{w_f}}{\sum_{i=0}^{C} \prod_{f=1}^{F} P_f(c_i|x)^{w_f}}.$$

4.2.2.4 Experiments and Results

The method was tested using the same data as in Sect. 4.2.1. The networks used consisted of two hidden layers with $2d$ nodes per hidden layer, where d is the size of the input vector.

The method was run for each of the metrics using all four feature-sets described in Sect. 4.2.1.2, but varying the total number of feature-sets from one to four in different runs. When less than four of the feature-sets were used they were selected in order of estimated utility (i.e. when three were used, the feature-set with the lowest estimated utility was ignored).

These results are shown in Table 4.5. When all four feature-sets were used the results were the same, but with reduction in the number of features an increasing improvement can be seen in comparison to the use of fixed sets of features. In Table 4.4, the highest performance for a single feature-set was 0.7212, but by selecting feature-sets on a leaf-by-leaf basis an accuracy of 0.8037 was achieved. The fourth metric, based on entropy, performed best and seems the natural choice since it is directly linked to the predictability of the result for a given part of the feature space.

Table 4.6 shows the frequency that each combination of features was used and the corresponding accuracies. Thus, for two feature-set cases, the combination of

Table 4.5 Classification accuracy for combinations of feature-sets and metrics

Metric	No. feature-sets			
	1	2	3	4
1	0.7575	0.8938	0.9406	0.9681
2	0.7738	0.8994	0.9275	0.9681
3	0.7763	0.9150	0.9456	0.9681
4	0.8037	0.9231	0.9538	0.9681

Table 4.6 Frequency and accuracy of each combination of feature sets

Shape	EFD	Blade	Margin	Acc	Frequency	P_{true}	P_{false}
✓				0.9080	0.0544	0.4077	0.2887
	✓			0.8437	0.2719	0.4169	0.2817
		✓		0.8186	0.1412	0.4013	0.3048
			✓	0.7688	0.5325	0.3982	0.2678
✓	✓			0.9744	0.0487	0.9186	0.5677
✓		✓		1.0000	0.0144	0.8583	0.0000
✓			✓	0.9682	0.0981	0.9198	0.4298
	✓	✓		0.9746	0.0737	0.9044	0.4208
	✓		✓	0.9038	0.5525	0.8845	0.5401
		✓	✓	0.9176	0.2125	0.8502	0.4058
✓	✓	✓		0.9600	0.0312	0.9656	0.5826
✓	✓		✓	0.9383	0.3950	0.9702	0.5743
✓		✓	✓	0.9851	0.0419	0.9605	0.3027
	✓	✓	✓	0.9624	0.5319	0.9514	0.5593
✓	✓	✓	✓	0.9681	1.0000	0.9779	0.6706

EFD and margin was used for 55.25% of the leaves. As would be expected, for most combinations the accuracy is higher than when applied to all the leaves in the test set, with those combinations that perform better in general being used with higher frequency. The combination of shape features and blade achieved 100% accuracy, but was only selected for a very small number of leaves. The final two columns, P_{true} and P_{false} show the average Fukunaga probability (used to determine the classification) when the result is correct and false respectively. The Fukunaga probability can be seen as a measure of confidence that a leaf has been correctly classified. This value is typically far higher when the classification is correct, suggesting it could be used to decide when to stop increasing the number of feature-sets used, since adding more tends to increase the Fukunaga probability.

In Table 4.7, the results are shown for the weighted and unweighted methods of selecting features and for a method using the same features for each leaf, with the order of application predetermined, based on their individual performances (in this case, the order is margin (accuracy 0.7212), shape features (0.6643), EFDs (0.6593) and texture (0.6437)). Both the weighted and unweighted methods performed significantly better than the fixed method when a reduced number of features was used. When more than one feature-set was used, the weighted version performed slightly better than the unweighted.

Finally, we examine whether the number of feature-sets required for the accurate classification of a leaf can be determined from the calculated Fukunaga probability. This could allow reduction to a minimum of the number of feature-sets used. The threshold of the probability was varied from 0 to 0.9, with additional feature-sets

Table 4.7 Results for the three methods of selecting features

Mode	No. features			
	1	2	3	4
Fixed	0.7212	0.8981	0.9025	0.9681
Unweighted	0.8037	0.9231	0.9538	0.9681
Weighted	0.8037	0.9256	0.9575	0.9700

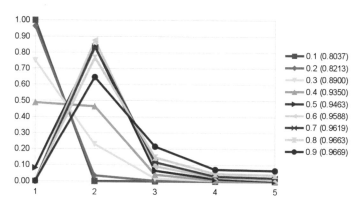

Fig. 4.6 Frequency (y-axis) of number of feature-sets (x-axis) required to meet Fukunaga probability threshold, unweighted case. The number in brackets indicates the accuracy achieved at that threshold. The x-axis value 5 indicates that the threshold was not met when using all features

being added until this threshold was passed. Once the threshold was passed the leaf was classified as a member of the species with the highest probability. Figures 4.6 and 4.7 show the frequency for which each number of feature-sets is required at each threshold, for the unweighted and weighted forms respectively. The x-axis value 5 indicates that the threshold was not met even when using all four features-sets. In most cases, only two feature-sets were required to meet this target. In all cases the modal number of features needed before the maximum accuracy was reached was two or less. In the weighted case, higher thresholds (above 0.5) required more features to be used, but did not improve accuracy.

Figure 4.8 shows the accuracy achieved at each threshold in relation to the average number of features required. The accuracy increases as the average number of feature-sets is increased, but there is little improvement in the results beyond two feature-sets.

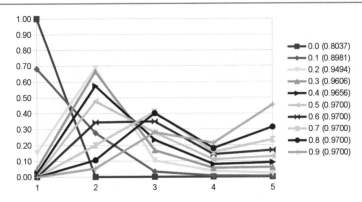

Fig. 4.7 Frequency of number of features required to meet Fukunaga probability threshold, weighted case

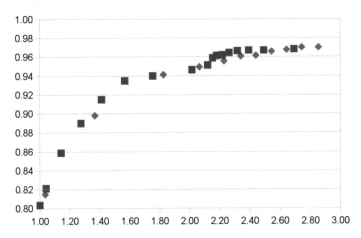

Fig. 4.8 Accuracy (y-axis) when different average numbers of feature-sets (x-axis) are used. *Blue* - unweighted; *Orange* - weighted

In the weighted case the maximum accuracy required an average of three feature-sets rather than four.

4.3   Summary

This chapter introduced a number of machine learning algorithms suitable for use in plant leaf classification. One of the biggest challenges involved in working with plants is the high level of within-species variation in morphological features and the considerable similarity in the leaves of some sets of species. Algorithms have been described here which increase the classification accuracy based on macro-texture and margin signatures by incorporating within-species variation present in the samples.

A framework has also been provided deriving the classification from combinations of different information sources. This includes the leaf-dependent selection of suitable feature-sets, in order to improve the classification whilst simultaneously decreasing computational costs and automatically eliminating detrimental feature-sets.

There is still much work that could be done in this area. As regards the incorporation of intra-species variation into the classification process, it would be useful to have such a method based on leaf shape, since this may vary considerably, even between leaves on the same plant. There are several further avenues worth exploring for the selection and combination of feature-sets. One is to investigate if the utility of a leaf component can be reliably estimated using simplified descriptors that are faster to extract. Elimination of descriptors of low utility would reduce the costs associated with compiling the more detailed descriptors used here. Another avenue is to include in the decision-making process the time required to develop and use each feature-set, since it may prove more efficient to use two fast-to-compare feature-sets than one slower but more reliable one.

Botanists' Vision

5

Chapter contributions:

- Study of the difference in eye movements between botanists and non-botanists when viewing leaf images, using eye-tracker data.
- Preliminary work towards replicating a botanist's observation points based on this data.

First-rate botanists have accumulated extensive experience in studying and identifying plants and this forms the foundation for their greater intuitive powers of species recognition. This suggests that it may be possible to improve on computational methods by investigating how botanists gather classificatory information from visual scanning of leaves, either from images or from actual specimens. Such data needs to be captured through the use of eye-tracking technology, since for the botanist this process is mostly intuitive.

When viewing any detailed image, such as an advertisement, website or some particular object, the attention of the human visual system is attracted to certain features, known as salient regions. This observation process is largely subconscious, although it can become less so through prior knowledge of the observed image, or experience in viewing particular types of image. Several research fields include work on eye movement, including and beyond the study of perceptual systems. The investigation of eye fixation points and saccades (fast eye movements between points of interest/stimuli) can provide insight into cognitive processes such as written language comprehension, memory, mental imagery and decision-making (Renniger et al. 2007). Eye movement research is important for neuroscience and psychiatry as well as ergonomics, advertising and design (Wedel and Pieters 2008). Since eye movements can be controlled voluntarily to some degree and detected and recorded by modern technology with great speed and precision, they can now be used as a powerful input device for many practical applications in human-computer interactions (Richardson et al. 2004).

© Springer-Verlag GmbH Germany 2017
P. Remagnino et al., *Computational Botany*, DOI 10.1007/978-3-662-53745-9_5

Wearable eye-tracking devices allow collection of information on eye-movement when looking at natural scenes, involving the use of generally unconstrained eye, head, and hand movements. The most commonly sought eye-tracking metrics include the number, duration and location of fixations, both across the entire scene and within set areas of interest, and the sequence of movements between them – there are many other such metrics (Megaw and Richardson 1979; Jacob and Karn 2003). Longer fixation periods generally indicate greater cognitive processing of the fixated region, possibly due to a higher level of detail or a lower scale feature of interest, and the percentage of total fixation dedicated to a particular area may indicate its saliency (Duchowski 2007; Ryan et al. 2010).

With sufficient knowledge and experience an expert in a particular field can become highly efficient at analysing certain types of images. This could be a physician searching for anomalies in images produced by medical scanners, a botanist studying images or specimens of leaves to determine a plant's species or security personnel identifying suspicious behaviour in CCTV footage. Using advanced eye-tracking technology, we can capture and analyse in great depth the process by which a human expert analyses such images. This chiefly involves identifying their fixations and analysing the sequence in which they are visited. Using this approach it may be possible to enable a computer system to replicate the human expert's fixation process more accurately. This in turn could lead to advances in the use of computer vision techniques to perform such tasks, since more efficient processing of the images would become possible and additional information overlooked by current techniques may be revealed. As regards plant identification, this could help to focus attention on those parts of the leaf that are most important in their classification into species or could inspire new processes yet to be considered.

In this chapter, eye-tracker data is used to perform preliminary work towards understanding how botanists study leaf images and for replicating the observation points a botanist uses when performing a leaf recognition task.

5.1 Comparing the Eye Movements of Botanists and Non-Botanists

Before eye-tracker data can be utilized, it is important to establish that the knowledge and experience acquired by botanists does indeed have an effect on their fixation points and the sequences thereof, when observing leaves. Here, a pilot study was conducted to demonstrate the difference in eye-movements between botanists and non-botanists. In the process, initial data was also gathered for use in replicating botanists' observations.

The experiments involve subjects carrying out a simple leaf recognition task. Subjects are shown an image of a leaf for a short period of time. Afterwards, they are allowed to view images of leaves from eight different species, one of which is from the same species as the initial leaf, and the subject is then asked to identify the correct leaf. Sets of species are chosen which have leaves portraying similar visual qualities (for example, similar coloration or shape), thereby making the task more difficult. This experiment was carried out at two different display intervals for the initial images, 1500 and 4000 ms. Each of these intervals was used eight times, using a different set of leaves for each replication.

During this task, the subject wore a head-mounted eye-tracking device. This has a camera capturing the view in front of the subject, a second camera capturing a video of one of their eyes and software capable of calculating from the eye-movements precisely where in their field of view they were looking. These observation points are recorded and translated back onto the original leaf image. Before each set of tests, the tracker was calibrated and the subject asked to minimise head movement during the test. Between each set of the tests, the subjects were allowed a short period of rest. The task was performed by nine volunteers – four botanists and five non-botanists.

5.1.1   Results and Analysis

The heatmaps – visualisations showing a subject's fixation points on an image – from a selection of the leaf images used are shown in Figs. 5.1 and 5.2, for the 1500 and 4000 ms cases respectively. The red spots represent the subjects' fixation points and those which are brightest red show the regions where their gaze remained for the longest time. The spots were generated as a Gaussian mixture model of each point the eye-tracker detected the subject looking at. They are normalised such that the point on which the subject's eyes spent the most time is shown as pure red. It can be seen that in the 1500 ms case the botanists typically viewed smaller, more select regions of the leaves than the non-botanists. This effect was diminished in the 4000 ms case when more time was available to look over the whole leaf. Another difference is that the focal points of the non-botanists more often lay beyond the leaf margin, suggesting they were paying more attention to details of leaf shape than to interior characteristics.

Figure 5.3 shows histograms of the fixation lengths - the length of time spent fixating on a single point - for both groups. The botanists' fixations tended to be longer (average 0.272 s vs. 0.209 s at 1500 ms, and 0.353 s vs. 0.284 s at 4000 ms) indicating that they observed areas in greater detail. In the 1500 ms tests, half the fixation time of the non-botanists lay in the 0.12 s interval, indicating that they did not concentrate for very long on any one part of the leaf, instead opting to scan a larger portion of the leaf in the time permitted.

The saccade amplitudes (angular magnitude of eye-movement) are shown in Fig. 5.4. Here there appears to be little significant difference between the botanists and non-botanists. The largest fraction of saccades in all cases was around 3°, equating to approximately one tenth of the leaf length, indicating a tendency to move from one region to another relatively close by, rather than hopping from one side of the leaf to the other.

Table 5.1 gives the average fixation lengths, saccade amplitudes and densities for each of the subjects. The density refers to a measure of the portion of a leaf viewed, with lower scores indicating that the subject concentrated on a smaller area of the

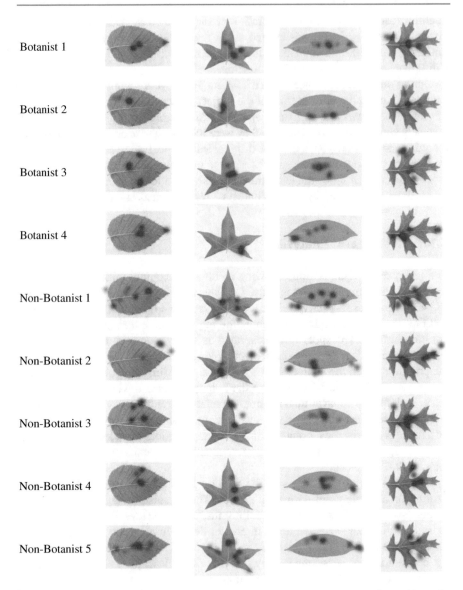

Fig. 5.1 Example heatmaps showing fixation points when leaves were exposed to subjects for 1500 ms

leaf. The botanists' average fixation lengths were greater than all but one of the non-botanists at 1500 ms, and longer than all of them at 4000 ms. For saccades, there was quite a wide variation between individuals, with average amplitudes ranging from $3.954°$ to $6.642°$, but no particular trend between the groups. In regard to density, when the viewing time was limited to 1500 ms all botanists had lower scores, but

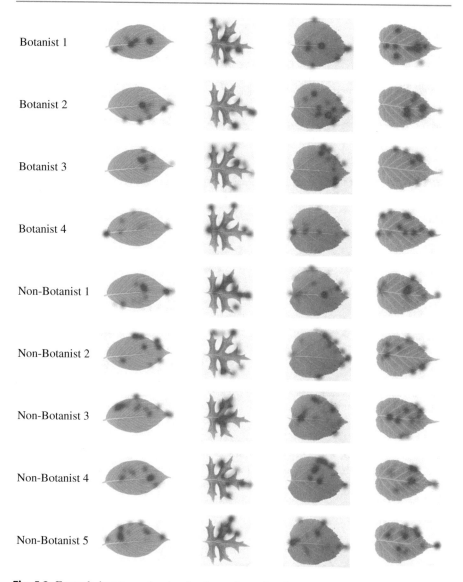

Fig. 5.2 Example heatmaps showing fixation points when leaves were exposed to subjects for 4000 ms

when the time was increased to 4000 ms, although all densities increased, two of the botanists had densities similar to those of the non-botanists, whilst the other two remained significantly lower, demonstrating variation in the botanists' behaviour.

This initial data appears to confirm the existence of a difference between the way experienced botanists and non-botanists view leaf images during a recognition task. The main quantifiable difference is that botanists preferred to study small parts of the

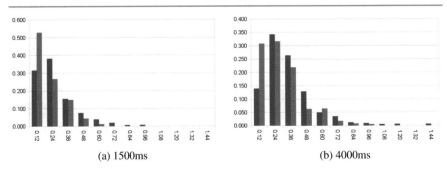

Fig. 5.3 Histogram of fixation lengths for botanists (*blue*) and non-botanists (*red*) for the 1500 and 4000 ms cases, showing length (x-axis, in seconds) against frequency (y-axis).

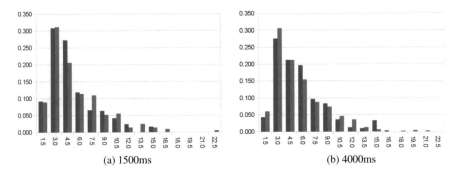

Fig. 5.4 Histogram of saccade amplitudes for botanists (*blue*) and non-botanists (*red*) for the 1500 and 4000 ms cases, showing amplitude (x-axis, in degrees) against frequency (y-axis).

leaf in greater detail, while the non-botanists attempted to acquire information from a larger portion of the leaf and consequently in less detail. There is also an indication that the non-botanists relied more on shape-related features than on features within the leaf contour.

5.2   Reverse Engineering of Expert Visual Observations

We present here a first step towards using this type of eye-tracking data for computer vision purposes, concentrating on its use in the study of the classification of plant leaves. We use the perspective of an expert in plant systematics, a fundamental field of plant biology in which tools based on morphology are used for identification purposes. Plant systematists are responsible for the organisation and accessibility of plant diversity data underpinned by accurate identification and naming. Figure 5.5

Table 5.1 Eye-tracking of leaves by botanists and non-botanists. Average statistics for each subject

Subject	Fixations		Saccades		Densities	
	1500 ms	4000 ms	1500 ms	4000 ms	1500 ms	4000 ms
Botanist 1	0.249	0.327	5.495	6.492	17218.0	35174.4
Botanist 2	0.301	0.338	4.557	4.890	17660.8	35174.4
Botanist 3	0.254	0.348	4.293	5.516	18585.1	23744.1
Botanist 4	0.282	0.399	5.714	5.910	18401.2	26539.0
Non-Botanist 1	0.206	0.262	6.254	6.072	25323.6	34644.9
Non-Botanist 2	0.168	0.280	6.642	6.252	22752.0	36157.9
Non-Botanist 3	0.194	0.263	3.954	4.668	21155.7	34299.7
Non-Botanist 4	0.261	0.319	5.427	5.762	20735.3	33704.9
Non-Botanist 5	0.218	0.295	5.105	4.913	22283.4	33422.8

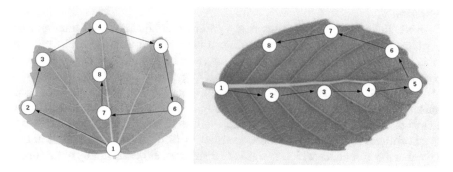

Fig. 5.5 Synthetic examples of typical sequences of fixation for eye-tracker data from an expert botanist

illustrates the typical sequences of fixations when an expert in plant systematics studies a leaf. In the approach used here, neural-gas algorithms (Martinetz and Shulten 1991) are applied for filter parameter learning to discover a set of filters which are particularly well suited for identifying the fixation points in a leaf image.

5.2.1 Related Work

In filter parameter learning (FPL) (Biem and Katagiri 1994; Heidemann 1996; Kurosawa 2008), a set of image filters is described by parameters whose values change dynamically through the course of some learning process. There have been numerous approaches to this problem. Heidemann (1996) presents an object recognition architecture based on feature extraction by Gabor filter kernels and performs feature classification using an artificial neural network. The parameters of the Gabor filters are optimized to the specific problem by minimizing an energy function. These filters can then be used to extract features that can be more easily classified by a neural network. Biem and Katagiri (1994) used a discriminative feature extraction method applied to a bank of filters for the modelling of speech. A method proposed by Kavukcuoglu et al. (2009) automatically learns the feature extractors in an unsupervised fashion by simultaneously learning the filters and the pooling units that combine multiple filter outputs. The method generates topographic maps of similar filters that extract features of orientations, scales and positions. In this way locally-invariant outputs are produced. Gautama and Hulle (1999) force the filters to partition the input space in an equitable manner: each filter is tuned to a different frequency region and contributes equally to the extraction of localized features. Here, a set of Gabor filters is learnt for processing images, due to their well-known ability to extract features from the parameters of frequencies, orientations and smoothing of the Gaussian envelope (Randen and Husoy 1999; Grigorescu et al. 2002; Chi et al. 2003; Li et al. 2010). Furthermore, links have been identified between Gabor filters and the human visual system (Daugman 1985) and thus may be beneficial for our purposes.

In the field of neural networks many different architectures and training rules exist, including perceptrons (from single-unit to multilayer versions), Hopfield-type recurrent networks (including probabilistic versions strongly related to statistical physics and Gibbs distributions) and the Self Organizing Map (SOM), among others (Feldkamp 1996; Haykin 2009). In a self-organising map, the network being trained has a fixed topology throughout. However there exist several variants where elements of the network are added or removed based on errors within the network. The neural-gas algorithm (Martinetz and Shulten 1991) is one such variant. It uses a fixed number of nodes which are initially distributed either randomly or uniformly throughout the input space. Connections between these nodes/neurons are added or removed so that for every input pattern, the two closest nodes are connected in the final network. In short, the organization of neurons, according to their distance to the input pattern and subsequent modification of their reference vector, produces the neuron expansion within the input space. The neurons' positions and their connections become configured so as to accurately represent the data distribution. Subsequently, by adding and deleting edges, a triangulation between different processing elements is produced. An extension to this, the growing-neural-gas algorithm (GNG) (Fritzke 1995), is initialised with just two nodes and adds more over time. It also deletes any nodes which have become separated from the network in an unused area of the space. This removes the requirement for a priori knowledge about the topological dimension of the space of input vectors (Rego et al. 2010). The method we use here is based on this form of the neural-gas algorithm.

5.2.2 Methodology

The approach described here finds a set of image filters that can be used to identify fixation points on an image of a plant leaf as efficiently as possible. First, data was collected on the location of such fixation points by capturing a botanist's eye movements as they studied a series of leaf images using an eye-tracking device. Each leaf was shown to the botanist for a set period of time, during which they were asked to provide as much verbal information as possible about the leaf. This was not recorded, but was done to ensure that the way in which the leaves were observed was realistic and relevant, as it meant the botanist had to look at areas of the leaf which would provide the most useful information. The fixation points discovered were used as input to an algorithm which attempts to find a set of filters that give high responses to fixation windows (Sect. 5.2.2.1). The filters thus learnt are based on the Gabor model (Sect. 3.2.2.1). The learning was performed using a variant of the growing neural gas algorithm (Sect. 5.2.2.3).

5.2.2.1 Fixations and Filter Responses

A *fixation point* is defined as a point in an image where a person focuses their attention for a short time (typically more than 100 ms, Blignaut 2009). The locations of these fixation points can be identified using eye-tracking technology when a series of images are shown to a subject. If each image is shown for only a very short time (no more than a few seconds), the fixation points found may correspond to the most salient parts of the image (Mannan et al. 2009). If, however, the expert is allowed to study an image for a longer time, the fixation points discovered will indicate the most important parts of the image needed by the expert for identification. Furthermore, the time that the expert spends concentrating on each fixation point and the order in which they move their vision between them can provide important information and insight into the processes being used by the experts.

In order to find and analyse fixation points, a *fixation window* is defined as a square region centred around a fixation point. The size of this window should correspond to the scale of the feature which the expert is studying, which may depend on the time which the expert spends looking at that feature. We chose to fix it at 100 pixels in width.

The method described here is intended for the discovery of filters that will be used for identifying fixation points. To achieve this, the response for a particular filter being applied to a particular fixation point is calculated as the sum of the absolute values of the convolution between filter and image at each pixel within the fixation window. The algorithm will search for a set of filters which produce high responses to fixation points.

5.2.2.2 Gabor Filters

The aim was to find a set of n filters, $F = \{f_1, f_2, ..f_n\}$, that could be used to identify fixations efficiently. For this purpose, Gabor filters (described in Sect. 3.2.2.1) were chosen. They have been applied to a large range of computer vision problems including image segmentation (Sandler and Lindenbaum 2006) and face detection (Huang et al. 2005). Gabor filters have been used in models of the human visual system and are therefore expected to prove useful here (Daugman 1980).

It is possible to produce a wide variety of different filters using only a small set
of parameters. The following parameter ranges were used: $\theta \in [0, \frac{\pi}{2}]$, $\gamma = 0.6$,
$\sigma \in [1, 10]$, $\lambda \in [\sigma, 8\sigma]$ and $\psi \in \{0, \frac{\pi}{2}\}$.

5.2.2.3 Learning the Filters

A neural gas algorithm was used to learn the set of filters. Fixations were chosen
one at a time at random from the training set and a filter was found that gave a
high response for that fixation. This filter was then used as an input pattern for the
neural gas algorithm. Filters were selected by testing a set number of filters sampled
from the portion of the input space occupied by the neural gas (see Algorithm 1
for details). The one with the highest response was chosen. This procedure avoids
wasting computation time and helps speed convergence by only testing filters that
are likely to prove useful.

The particular neural gas algorithm used here is a modification of the growing
neural gas algorithm (Fritzke 1995). The original algorithm was initialised with two
neurons and grown by adding a new neuron every set number of iterations. Since we
wish to find only a minimal number of filters (using a large number of filters would
reduce efficiency when searching for fixations), the algorithm instead starts with the
maximum required number of neurons, and only adds new ones if the algorithm
removes a neuron which has become separated from the network, and so the number
of neurons remains constant. The advantage of using this approach over the standard
neural gas algorithm is that it replaces neurons that appear less useful, thereby aiding
convergence. Furthermore, the removal of connections allows the gas to separate if
discrete regions of the space need to be occupied. Once the algorithm has converged,
post-pruning (Canales and Chacón 2007) is applied to further improve the final set
of filters. The post-pruning algorithm removes clusters of neurons from unused parts
of the space and adjusts the positions of others to achieve better final results. The
algorithm can be summarised as follows:

1. Initialise the neural gas:

    - Neurons are uniformly distributed through filter-space within predetermined
      bounds for each parameter.
    - Connections are created between neighbouring neurons. Each connection has
      an age initially set to 0.

2. Until stopping criteria are met, repeat:

    a. Select a fixation from the training set.
    b. Generate k filters drawn from the distribution of neurons as in Algorithm 1;
        this generates random filters, returning the first which matches the criteria.
    c. Calculate the response for each of the filters being applied to the training
        fixation.

    d. Apply one step of the neural gas algorithm (see section "A Modified Growing
       Neural Gas Algorithm"), using as the input pattern the parameters of the filter
       with the highest response in the previous step.

3. Apply post-pruning (Canales and Chacón 2007) to the final neural gas.

Algorithm 1 Kernel density estimation algorithm for selecting filters

---

repeat
    $\xi \leftarrow$ random filter vector
    $x \leftarrow 0$
    for all neurons $f_i \in F$ **do**
        $x \leftarrow x + \dfrac{1}{\sqrt{2\pi\sigma^2}} \exp^{-\frac{||\xi - f_i||^2}{2\sigma^2}}$
    end for
    $y \leftarrow$ random value in range $[0, \frac{|F|}{\sqrt{2\pi\sigma^2}}]$
until $y \leq x$
return ξ

A Modified Growing Neural Gas Algorithm

At each step the modified growing neural gas algorithm calculates the new positions
of its neurons according to an input pattern ξ:

1. Find the two neurons, f_1, f_2 closest (by Euclidean distance) to the input pattern
   ξ.
2. Increment the age of all connections between f_1 and its neighbours.
3. Increase f_1's accumulated error by $||f_1 - \xi||^2$.
4. Move f_1 and its connected neurons towards ξ, by fractions ε_b, ε_c (here 0.2 and
   0.1) respectively:

$$f_1 = f_1 + \varepsilon_b(\xi - f_1)$$
$$f_c = f_c + \varepsilon_c(\xi - f_c) \text{ for all direct neighbours } c \text{ of } f_1$$

5. If no connection exists between f_1, f_2, create a new connection, else reset the
   connections age to 0.
6. Remove any connection with an age above some threshold.
7. Remove any neurons which have become disconnected from all other neurons.
8. For each neuron removed in step 7, insert a new neuron as follows:

    a. Find the neuron, f_i, with the largest accumulated error (from step 3), and the
       neuron, f_j, with the highest accumulated error of all f_i's neighbours.
    b. Insert a new neuron, f_k, between f_i and f_j:

$$f_k = \frac{f_i - f_k}{2}$$

    c. Replace the connection between f_i, f_j with new connections between f_i, f_k
      and f_j, f_k.

Fig. 5.6 Heatmaps from eye-tracker data indicating the leaf insertion points and apices as fixations

    d. Decrease the accumulated errors of f_i, f_j, by multiplying them by some constant. Set accumulated error of f_k equal to that of f_i.

9. Decrease all error variables by multiplying them by some constant.

5.2.3 Evaluation

Due to the nature of the data acquired by an eye-tracker, quantitative results are difficult to obtain. This is because there are no definitive negative testing examples. For example, when studying a leaf, a botanist may only need to look at the margin on one side of the leaf to obtain the information he requires from it. If the margin on the other side of the leaf is not viewed, this does not mean it is any less relevant, since the decision to use one particular side may have been arbitrary due to leaf symmetry and may be different on a second viewing of the same leaf. Because of this, instead of trying to identify all possible fixation points on a leaf image, the evaluation method instead tries to locate a few different leaf features and treat all other areas of the leaf as negative examples.

The data from expert botanists showed that the insertion point of the leaf (where the petiole or leaf stalk joins the leaf) and the apex (the tip of the leaf) are fixation points on most leaves (Fig. 5.6). The filters which have been learnt are used to identify these fixations from a set of points randomly taken from some leaf images. In the first experiment (section "Experiment 1") we used a nearest-neighbour classifier to label image windows as either insertion point, apex or other. In the second (section "Experiment 2") the points on each leaf most likely to be apex and insertion point were found. The results were compared to those using filters similar to the popular Leung-Malik and Root Filter Set filter banks (Leung and Malik 2001; Geusebroek et al. 2002). The results are discussed in section "Results".

The images used were oriented automatically according to the primary axis of the leaf and scaled so that each leaf had an area within the image of approximately 2^{18} pixels. A fixation window width of 100 pixels was used, as this allowed for an appropriate size region around the apices and insertion points.

Experiment 1

The first experiment analysed whether the filters learnt for identifying fixation windows were suitable for this purpose. The training set consisted of the windows cen-

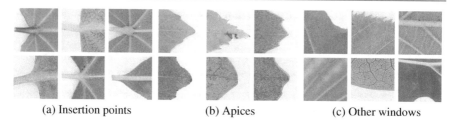

(a) Insertion points   (b) Apices   (c) Other windows

Fig. 5.7 Examples of the three fixation classes

Table 5.2 Results for experiment 1 (% windows correctly classified)

Method	16 filters	36 filters	128 filters
New method	93.68	93.26	93.61
Filter bank	80.97	89.58	93.68

tred around the insertion points and apices of 240 leaves from a total of 30 different species. The testing set then comprised windows centred around the insertion point, apex and four other points on a different set of 240 leaves (see Fig. 5.7 for some examples). Once a set of filters had been learnt using the training set, they were used to process each fixation in the training set to produce an n-dimensional vector of filter responses that could be used to describe the fixations. Filter response vectors were then generated for the fixations in the test set. A nearest-neighbour algorithm was used to classify the test fixations, thus finding the nearest training vector to each test vector. If the distance between them was less than some threshold, the test fixation was assigned to the class of the training fixation, if greater than the threshold then it was excluded from that class.

For comparison, a set of filters based on the Leung-Malik and Root Filter Set filter banks was also used. These filters are evenly distributed throughout the parameter space, using the same parameter ranges as for the new method. The results with 16, 36 and 128 filters are given in Table 5.2, the values of which indicate the percentage of windows correctly classified. The ROC curve shown in Fig. 5.8 was produced by varying the threshold used to perform the classification, for the sets of 16 filters. The accuracies given in Table 5.2 correspond to the threshold value which gave the best results for each set of filters.

Experiment 2

In our second experiment, we attempted to find the insertion point and apex on a leaf from among 16 possible windows. This experiment more accurately reflects the final task, as it attempted to discover the most likely fixation points on each image. The training was performed as in experiment 1. Each leaf image was tested by comparing the response vectors for windows centred around the insertion point, apex and 14 other randomly selected points to the response vectors for the training

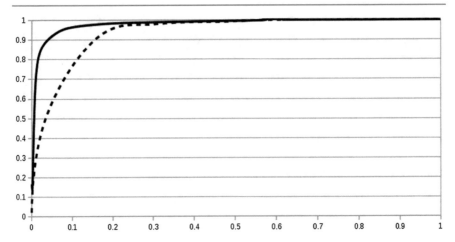

Fig. 5.8 ROC curve for experiment 1. *Solid line* = our method, *dashed line* = standard filter bank

Table 5.3 Results for experiment 2 (% windows correctly classified)

Method	16 filters	36 filters	128 filters
Our method	91.25	95.00	95.08
Filter bank	77.92	91.76	94.58

set. The windows closest to an insertion point and apex from the training set were selected as being these respective points on the test leaf. Table 5.3 shows the results of this experiment.

Results

The results show that both sets of filters achieved a similarly high level of accuracy when using a large number of filters. The filters learnt using the algorithm retained their quality when their number was reduced to just 16, but the filters based on the standard filter banks performed significantly worse. In the case of the filter banks, filters were removed indiscriminately from useful areas of the parameter space when the total number of filters was reduced, causing loss of accuracy. In contrast, the filters learnt by our algorithm continued to perform well because they occupied only useful regions in the space. The large number of filters meant there was redundancy in these regions and reduction in filter number removed only the redundancy without affecting quality. The same level of accuracy was achieved with only a fraction of the processing.

5.2.4 Summary

For many practical applications speed and efficiency are particularly important for discovering the fixation points in an image, because large areas of the image need to be processed. Our algorithm is an advance in that it allows the learning of a small set of filters capable of distinguishing fixation points with an accuracy equal to that achieved by a larger filter bank. Further work in this area could explore how efficiency can be further improved by intelligently selecting only the best subset of our filters to use on any given part of the image. It would also be useful to investigate the use of different methods of classification for improving accuracy.

Beyond this, methods for the discovery of the fixation sequences need to be investigated. One possibility may be the use of hidden Markov models, involving spatial and temporal information, as well as data generated by the filter set. With a system for accurate estimation and replication of the methods by which a human expert studies an image, it may be possible to improve the efficiency and robustness of automated computer vision systems for performing such tasks. However, investigation is needed to evaluate whether such information does indeed lead to more accurate leaf classification. In addition, we may be able to achieve a better understanding of how human experts carry out this task.

References

D.C. Adams, F.J. Rohlf, D.E. Slice, Geometric morphometrics: ten years of progress following the "revolution". Italian J. Zool. **71**(1), 5–16 (2004)

G. Agarwal, P. Belhumeur, S. Feiner, D. Jacobs, W.J. Kress, R. Ramamoorthi, N.A. Bourg, N. Dixit, H. Ling, D. Mahajan, First steps toward an electronic field guide for plants. Taxon **55**(3), 597 (2006)

E. Anderson, The species problem in Iris. Ann. Mo. Bot. Gard. **23**, 457–509 (1936a)

E. Anderson, Hybridization in American Tradescantias. Ann. Mo. Bot. Gard. **23**, 511–525 (1936b)

E. Anderson, The technique and use of mass collections in plant taxonomy. Ann. Mo. Bot. Gard. **28**, 287–292 (1941)

E. Anderson, Maize in Mexico: a preliminary survey. Ann. Mo. Bot. Gard. **33**, 147–247 (1946)

E. Anderson, *Introgressive Hybridization* (Wiley, New York, 1949)

E. Anderson, L.N. Abbe, A quantitative comparison of specific and generic differences in the Betulaceae. J. Arnold Arbor. **15**, 43–49 (1934)

E. Anderson, W.B. Turrill, Biometrical studies on herbarium material. Nature **136**, 986–987 (1935)

E. Anderson, T.W. Whitaker, Speciation in Uvularia. J. Arnold. Arbor. **15**, 28–42 (1934)

I.M. Andrade, S.J. Mayo, D. Kirkup, C. Van Den Berg, Comparative morphology of populations of Monstera Adans. (Araceae) from natural forest fragments in northeast Brazil using elliptic Fourier analysis of leaf outlines. Kew Bull. **63**, 193–211 (2008)

I.M. Andrade, S.J. Mayo, D. Van den Berg, C. Kirkup, Elliptic Fourier analysis of leaf shape in Anthurium sinuatum Benth. ex Schott and A. pentaphyllum (Aubl.) G. Don (Araceae) in forest fragment populations from Northeast Brazil. Kew Bull. **65**, 3–20 (2010)

E. Ashby, Studies in the morphogenesis of leaves I. An essay on leaf shape. N. Phytol. **47**(2), 153–176 (1948)

A.F. Atiya, Estimating the posterior probabilities using the k-nearest neighbor rule. Neural Comput. **17**(3), 731–740 (2005)

S. Atran, *Cognitive Foundations of Natural History: Towards an Anthropology of Science* (Cambridge University Press, Cambridge, 1990)

A.R. Backes, O.M. Bruno, Plant leaf identification using multi-scale fractal dimension, in *International Conference On Image Analysis and Processing* (Springer, Heidelberg, 2009), pp. 143–150

© Springer-Verlag GmbH Germany 2017
P. Remagnino et al., *Computational Botany*, DOI 10.1007/978-3-662-53745-9

A.R. Backes, J.J.M. Sá Jr., R.M. Kolb, O.M. Bruno, Plant species identification using multi-scale fractal dimension applied to images of adaxial surface epidermis. Comput. Anal. Images Patterns **5702**, 680–688 (2009)

A.R. Backes, W.N. Gonçalves, A.S. Martinez, O.M. Bruno, Texture analysis and classification using deterministic tourist walk. Pattern Recognit. **43**, 685–694 (2010)

J.F. Barczi, H. Rey, Y. Caraglio, P. de Reffye, D. Barthélemy, Q.X. Dong, T. Fourcaud, Mapsim: a structural whole-plant simulator based on botanical knowledge and designed to host external functional models. Ann. Botany **101**(8), 1125–1138 (2007)

D. Barthélemy, Y. Caraglio, Plant architecture: a dynamic, multilevel and comprehensive approach to plant form, structure and ontogeny. Ann. Botany **99**, 375–407 (2007)

P. Belhumeur, D. Chen, S. Feiner, D. Jacobs, W. Kress, H. Ling, I. Lopez, R. Ramamoorthi, S. Sheorey, S. White, L. Zhang, *Searching the World's Herbaria: A System for Visual Identification of Plant Species, in Computer Vision - ECCV 2008* (Springer, Heidelberg, 2008)

A.D. Bell, P.B. Tomlinson, Adaptive architecture in rhizomatous plants. Bot. J. Linn. Soc. **80**, 125–160 (1980)

S. Belongie, J. Malik, J. Puzicha, Shape matching and object recognition using shape contexts. IEEE Trans. Pattern Anal. Mach. Intell. 509–522 (2002)

W.G. Berendsohn, *More Tax - Handling Factual Information Linked to Taxonomic Concepts in Biology*, vol. 39, *Schriftenreihe für Vegetationskunde* (Bundesamt für Naturschutz, 2003)

A. Biem, S. Katagiri, Filter bank design based on discriminative feature extraction, in *IEEE International Conference on Acoustics, Speech, and Signal Processing* (1994), pp. 485–488

R.E. Blackith, Morphometrics, in *Theoretical and Mathematical Biology*, ed. T.H. Waterman, H.J. Morowitz (Blaisdell Publishing Company, 1965), pp. 225–249

R.E. Blackith, R.A. Reyment, *Multivariate Morphometrics* (Academic Press, London, 1971)

P. Blignaut, Fixation identification: The optimum threshold for a dispersion algorithm. Atten. Percept. Psychophys. **71**(4), 881–895 (2009)

V. Bonhomme, S. Picq, C. Gaucherel, J. Claude, Momocs: outline analysis using R. J. Stat. Softw. **56**(13), 1–24 (2014)

F.L. Bookstein, Foundations of morphometrics. Ann. Rev. Ecol. Syst. **13**, 451–470 (1982)

F.L. Bookstein, Size and shape spaces for landmark data in two dimensions. Stat. Sci. **1**(2), 181–222 (1986)

F.L. Bookstein, Principal warps: thin-plate splines and the decomposition of deformations. IEEE Trans. Pattern Anal. Mach. Intell. **11**(6), 567–585 (1989)

F.L. Bookstein, *Morphometric Tools for Landmark Data: Geometry and Biology* (Cambridge University Press, Cambridge, 1991)

F.L. Bookstein, A brief history of the morphometric synthesis, in *Contributions to Morphometrics*, Monografias del Museo Nacional de Ciencias Naturales (C.S.I.C., 1993), pp. 15–40

F.L. Bookstein, Landmark methods for forms without landmarks: morphometrics of group differences in outline shape. Med. Image Anal. **1**, 225–243 (1997)

W. Borkowski, Fractal dimension based features are useful descriptors of leaf complexity and shape. Canadian J. For. Res. **29**(9), 1301–1310 (1999)

D. Briggs, S.M. Walters, *Plant Variation and Evolution* (Cambridge University Press, Cambridge, 1997)

M. Budka, B. Gabrys, K. Musial, On accuracy of PDF divergence estimators and their applicability to representative data sampling. Entropy **13**, 1229–1266 (2011)

X.P. Burgos-Artizzu, A. Ribeiro, A. Tellaeche, G. Pajares, C. Fernández-Quintanilla, Analysis of natural images processing for the extraction of agricultural elements. Image Vis. Comput. **28**(1), 138–149 (2010)

K.M. Butterworth, D.C. Adams, H.T. Horner, J.F. Wendel, Initiation and early development of fiber in wild and cultivated cotton. Int. J. Plant Sci. **170**(5), 561–574 (2009)

A.J. Cain, The evolution of taxonomic principles, in *Microbial Classification*, ed. by G.C. Ainsworth, P.H.A. Sneath (Cambridge University Press, Cambridge, 1962), pp. 1–13

A.J. Cain, G.A. Harrison, Phyletic weighting. Proc. Zool. Soc. Lond. **135**, 1–31 (1960)

J.H. Camin, R.R. Sokal, A method for deducing branching sequences in phylogeny. Evolution **19**(3), 311–326 (1965)

F. Canales, M. Chacón, Modification of the growing neural gas algorithm for cluster analysis. Progr. Pattern Recogn. Image Anal. Appl. **4756**, 684–693 (2007)

C.H. Cannon, P.S. Manos, Combining and comparing morphometric shape descriptors with a molecular phylogeny: the case for fruit type evolution in bornean lithocarpus (fagaceae). Syst. Biol. **50**(6), 860–880 (2001)

A. Cardini, A. Loy, On growth and form in the computer era: from geometric to biological morphometrics. Hystrix: the Italian. J. Mammal. **24**(1), 1–5 (2013)

D. Casanova, J.J.M. Sá Jr., O.M. Bruno, Plant leaf identification using Gabor wavelets. Int. J. Imag. Syst. Technol. **19**, 236–243 (2009)

CBOL Plant Working Group, A DNA barcode for land plants. Proc. Natl Acad. Sci. **106**, 12794–12797 (2009)

K. Chatfield, V. Lempitsky, A. Vedaldi, A. Zisserman, The devil is in the details: an evaluation of recent feature encoding methods, in *British Machine Vision Conference* (2011), pp. 76.1–76.12

L. Chen, B. Chen, Y. Chen, Image feature selection based on ant colony optimization, in *Advances in Artificial Intelligence* (2011), pp. 580–589,

X. Chen, X. Hu, X. Shen, Spatial weighting for bag-of-visual-words and its application in content-based image retrieval. Adv. Knowl. Discov. Data Min. **5476**, 867–874 (2009)

S.C. Cheng, J.J. Jhou, B.H. Liou, *PDA Plant Search System Based on the Characteristics of Leaves Using Fuzzy Function, in New Trends in Applied Artificial Intelligence, number 4570 in LNAI* (Springer, Heidelberg, 2007)

Z. Chi, L. Houqiang, W. Chao, Plant species recognition based on bark patterns using novel Gabor filter banks, in *Proceedings of the 2003 International Conference on Neural Networks and Signal Processing, 2003*, vol. 2 (2003), pp. 1035–1038

D.H. Chitwood, A. Ranjan, C.C. Martinez, L.R. Headland, T. Thiem, R. Kumar, M.F. Covington, T. Hatcher, D.T. Naylor, S. Zimmerman, N. Downs, N. Raymundo, E.S. Buckler, J.N. Maloof, M. Aradhya, B. Prins, L. Li, S. Myles, N.R. Sinha, A modern ampelography: a genetic basis for leaf shape and venation patterning in grape. Plant Physiol. **164**, 259–272 (2014)

B.R. Clark, H.C.D. Godfray, I.J. Kitching, S.J. Mayo, M.J. Scoble, Taxonomy as an escience. Philos. Trans. R. Soc. A **367**, 953–966 (2009)

J.Y. Clark, Neural networks and cluster analysis for unsupervised classification of cultivated species of Tilia (Malvaceae). Bot. J. Linn. Soc. **159**, 300–314 (2009)

J. Clarke, S. Barman, P. Remagnino, K. Bailey, D. Kirkup, S. Mayo, P. Wilkin, Venation pattern analysis of leaf images. Lect. Notes Comput. Sci. **4292**, 427–436 (2006)

J. Clausen, Studies on the collective species of viola tricolor l. Botanisk Tidsskrift **37**, 363–416 (1922)

J. Clausen, D.D. Keck, W.M. Hiesey, *Experimental Studies on the Nature of Species I. The Effect of Varied Environments on Western North American Plants*, vol. 520 (Carnegie Institution of Washington, Washington D.C, 1940)

J. Clausen, D.D. Keck, W.M. Hiesey, *Experimental Studies on the Nature of Species, II. Plant Evolution Through Amphiploidy and Autoploidy, with Examples from the Madiinae*, vol. 564 (Carnegie Institution of Washington Publication, Washington D.C, 1945)

P. Comon, Independent component analysis: a new concept? Signal Process. **36**, 287–314 (1994)

D.P.A. Corney, J.Y. Clark, H.L. Tang, P. Wilkin, Automatic extraction of leaf characters from herbarium specimens. Taxon **61**(1), 231–244 (2012)

G. Csurka, C.R. Dance, L. Fan, J. Willamowski, C. Bray, Visual categorization with bags of keypoints, in *Workshop on Statistical Learning in Computer Vision, ECCV* (2004), pp. 1–22

H. Dale, A. Runions, D. Hobill, P. Prusinkiewicz, Modelling biomechanics of bark patterning in grasstrees. Ann. Botany **114**, 629–641 (2014)

C. Darwin, *On the Origin of Species by Means of Natural Selection, or the Preservation of Favoured Races in the Struggle for Life* (J. Murray, London, 1859)

M. Das, R. Manmatha, E.M. Riseman, Indexing flower patent images using domain knowledge. IEEE Intell. Syst. **14**(5), 24–33 (1999)

S. Dasgupta, The evolution of the D^2-statistic of Mahalanobis. Indian J. Pure Appl. Math. **26**(6), 485–501 (1995)

J.G. Daugman, Two-dimensional spectral analysis of cortical receptive field profiles. Vis. Res. **20**, 847–856 (1980)

J.G. Daugman, Uncertainty relation for resolution in space, spatial frequency, and orientation optimized by two-dimensional visual cortical filters. J. Opt. Soc. Am. A **2**, 1160–1169 (1985)

J.F. Davidson, The polygonal graph for simultaneous portrayal of several variables in population analyses. Madroño **9**, 105–110 (1947)

A.P. de Candolle, *Théorie élémentaire de la botanique; ou, Exposition des principes de la classification naturelle et de l'art de décrire et d'étudier les végétaux.* Déterville (1813)

T.A. Dickinson, W.H. Parker, R.E. Strauss, Another approach to leaf shape comparisons, in *Taxon* 1–20 (1987)

L.L. Dryden, K.V. Mardia, *Statistical Shape Analysis* (Wiley, Chichester, 1998)

J.-X. Du, D.-S. Huang, X.-F. Wang, X. Gu, Computer-aided plant species identification (CAPSI) based on leaf shape matching technique. Trans. Inst. Meas. Control **28**, 275–284 (2006)

J.-X. Du, X.-F. Wang, G.-J. Zhang, Leaf shape based plant species recognition. Appl. Math. Comput. **185**, 883–893 (2007)

A.T. Duchowski, *Eye Tracking Methodology: Theory and Practice* (Springer, New York Inc, Secaucus, 2007)

J. Duminil, D. Kenfack, V. Viscosi, L. Grumiau, O.J. Hardy, Testing species delimitation in sympatric species complexes: the case of an african tropical tree. Carapa spp. (Meliaceae). Mol. Phylogenet. Evol. **62**, 275–285 (2012)

E.M. East, Studies on size inheritance in Nicotiana. Genetics **1**, 164–176 (1915)

J. Edmonds, R.M. Karp, Theoretical improvements in algorithmic efficiency for network flow problems. J. ACM **19**, 248–264 (1972)

B. Ellis, D.C. Richardson, M.J. Spivey, W.G. Daly, L.J. Hickey, K.R. Johnson, J.D. Mitchell, P. Wilf, S.L. Wing, *Manual Of Leaf Architecture* (Cornell University Press, Ithaca, 2009)

C. Epling, *The Living Mosaic, Research Lecture* (University of California Press, California, 1944)

R.O. Erickson, Population size and geographical distribution of Clematis fremontii var. riehlii. Ann. Mo. Bot. Gard. **30**, 63–68 (1943)

R.O. Erickson, The Clematis fremontii var. riehlii population in the Ozarks. Ann. Mo. Bot. Gard. **32**, 413–460 (1945)

D.S. Falconer, *Introduction to Quantitative Genetics* (Longman Scientific and Technical, 1989)

J.S. Farris, The meaning of relationship and taxonomic procedure. Syst. Zool. **16**(1), 44–51 (1967)

N.C. Fassett, Mass collections: Rubus odoratus and R. parviflorus. Ann. Mo. Bot. Gard. **28**(3), 299–374 (1941)

N.C. Fassett, Mass collections: Diervilla lonicera. Bull. Torrey Bot. Club **69**(4), 317–322 (1942)

N.C. Fassett, The validity of Juniperus virginiana var. crebra. Am. J. Botany **30**(7), 469–477 (1943)

L.A. Feldkamp, Neural networks: Current applications, book review. Proc. IEEE **84**(1), 87 (1996)

R.A. Fisher, The correlation between relatives on the supposition of mendelian inheritance. Trans. R. Soc. Edinb. **52**, 399–433 (1918)

R.A. Fisher, The use of multiple measurements in taxonomic problems. Ann. Eugen. **7**(2), 179–188 (1936)

R.A. Fisher, *The Genetical Theory of Natural Selection* (Oxford University Press, Oxford, Variorum Edition, 1999)

D. Frijters, A. Lindenmayer, A model for the growth and flowering of Aster novae-angliae on the basis of table (1,0) L-systems, in *L-systems and Lecture Notes in Computer Science*, ed. by G. Rozenberg, A. Samolaa, vol. 15, (Springer, Berlin, 1974), pp. 24–52

B. Fritzke, *A growing neural gas network learns topologies, in Advances in Neural Information Processing Systems* (MIT Press, Cambridge, 1995), pp. 625–632

H. Fu, Z. Chi, Combined thresholding and neural network approach for vein pattern extraction from leaf images, in *IEEE Proceedings Vision Image and Signal Processing*, vol. 153 (Institution of Electrical Engineers, 2006), pp. 881–892

K. Fukunaga, L. Hostetler, K-nearest-neighbor Bayes-risk estimation. IEEE Trans. Inf. Theory **21**(3), 285–293 (1975)

N. Furuta, S. Ninomiya, N. Takahashi, H. Ohmori, Y. Ukai, Quantitative evaluation of soybean leaflet shape by principal component scores based on elliptic Fourier descriptor. Breed. Sci. **45**, 315–320 (1995)

K.R. Gabriel, R.R. Sokal, A new statistical approach to geographic variation analysis. Syst. Zool. **18**(3), 259–278 (1969)

E. Gage, P. Wilkin, A morphometric study of species delimitation in Sternbergia lutea (Alliaceae, Amaryllidoideae) and its allies S. sicula and S. greuteriana. Bot. J. Linn. Soc. **158**, 460–469 (2008)

F. Galton, *Natural Inheritance* (MacMillan and Co., London, 1889)

K.J. Gaston, M.A. O'Neill, Automated species identification: why not? Philos. Trans. R. Soc. Lond. B **359**(1444), 655–667 (2004)

T. Gautama, M.M. Van Hulle, Self-organized feature extraction achieved with a parameterized filterbank. Neural Proc. Lett. **10**(2), 131–137 (1999)

S. Gebhardt, J. Schellberg, R. Lock, W. Kühbach, Identification of broad-leaved dock (Rumex obtusifolius L.) on grassland by means of digital image processing. Precis. Agric. **7**(3), 165–178 (2006)

J. Geusebroek, A. Smeulders, J. van de Weijer, Fast anisotropic gauss filtering. IEEE Trans. Image Process. **12**, 2003 (2002)

H.C.J. Godfray, Challenges for taxonomy. Nature **2002**(417), 17–19 (2002)

J.M. Gómez, F. Perfectti, J.P.M. Camacho, Natural selection on Erysimum mediohispanicum flower shape: insights into the evolution of zygomorphy. Am. Nat. **168**(4), 531–545 (2006)

E.G. Gonçalves, H. Lorenzi, *Morfologia Vegetal: Organografia e dicionario ilustrado de morfologia das plantas vasculares* (Instituto Plantarum, Nova Odessa, 2011)

M.D. Gordin, *A Well-ordered Thing: Dmitrii Mendeleev and the Shadow of the Periodic Table* (Basic Books, New York, 2004)

A.D. Gordon, *Classification, in Monographs on Statistics and Applied Probability* (Chapman and Hall/CRC, London, 1999)

J.C. Gower, Some distance properties of latent root and vector methods used in multivariate analysis. Biometrika **53**(3/4), 325–338 (1966)

J.C. Gower, A general coefficient of similarity and some of its properties. Biometrics **27**(4), 857–871 (1971)

J.C. Gower, The biological stimulus to multidimensional analysis. Journal Electronique d'Histoire des Probabilités et de la Statistique **4**(2), 1–14 (2008)

J.W. Gregor, V.M. Davey, J.M.S. Lang, Experimental taxonomy I. Experimental garden technique in relation to the recognition of small taxonomic units. N. Phytol. **35**(4), 323–350 (1936)

S.E. Grigorescu, N. Petkov, P. Kruizinga, Comparison of texture features based on Gabor filters. IEEE Trans. Image Process. **11**(10), 1160–1167 (2002)

S. Gubatz, V.J. Dercksen, C. Brüss, W. Weschke, U. Wobus, Analysis of barley (Hordeum vulgare) grain development using three-dimensional digital models. Plant J. **52**(4), 779–790 (2007)

P. Gunz, P. Mitteroecker, Semilandmarks: a method for quantifying curves and surfaces. Hystrix **24**(1), 103–109 (2013)

A. Haigh, P. Wilkin, F. Rakotonasolo, A new species of Dioscorea L. (Dioscoreaceae) from Western Madagascar and its distribution and conservation status. Kew Bull. **60**, 273–281 (2005)

F. Hallé, R.A.A. Oldeman, *Essai sur l'architecture e la dynamique de croissance des arbres tropicaux* (Masson, Paris, 1970)

F. Hallé, R.A.A. Oldeman, P.B. Tomlinson, *Tropical Trees and Forests - An Architectural Analysis* (Springer, Heidelberg, 1978)

M.B. Hamilton, *Population Genetics* (Wiley-Blackwell, New York, 2009)

Ø. Hammer, *PAST Paleontological Statistics, in Natural History Museum* (University of Oslo, Norway, 2012)

Ø. Hammer, D.A.T. Harper, *Paleontological Data Analysis* (Wiley-Blackwell, New York, 2008)

Ø. Hammer, D.A.T. Harper, P.D. Ryan, Past: Paleontological statistics software package for education and data analysis. Palaeontologia Electronica **4**(1), 1–9 (2001)

R.M. Haralick, I. Dinstein, K. Shanmugam, Textural features for image classification. IEEE Trans. Syst. Man Cybern. SMC- **3**, 610–621 (1973)

J.L. Harper, *Population Biology of Plants* (Academic Press, London, 1977)

W. Hawthorne, C. Jongkind, *Woody Plants of Western African Forests: A Guide to the Forest Trees, Shrubs and Lianes from Senegal to Ghana* (Kew Publishing, Kew, Royal Botanic Gardens, 2006)

S. Haykin, *Neural Networks and Learning Machines* (Prentice Hall, Upper Saddle River, 2009)

D.J. Hearn, Shape analysis for the automated identification of plants from images of leaves. Taxon **58**, 934–954 (2009)

G. Heidemann, A neural 3-D object recognition architecture using optimized Gabor filters, in *Proceedings of the International Conference on Pattern Recognition (ICPR 96)* (IEEE Computer Society, Washington DC USA, 1996), pp. 70–74

C.B. Heiser, E. Anderson, Botanist and curator of useful plants. Ann. Mo. Bot. Gard. **82**(1), 54–60 (1995)

J. Hemming, T. Rath, PA - Precision agriculture: computer-Vision-based weed identification under field conditions using controlled lighting. J. Agricul. Eng. Res. **78**(3), 233–243 (2001)

A. Henderson, A revision of Geonoma (Arecaceae). Phytotaxa **17**, 1–271 (2011)

W. Hennig, *Phylogenetic Systematics* (University of Illinois Press, Urbana, 1966)

L.J. Hickey, Classification of the architecture of dicotyledonous leaves. Am. J. Botany **60**(1), 17–33 (1973)

H. Honda, Description of the form of trees by parameters of the tree-like body: effects of the branching angle and branch length on the shape of the tree-like body. J. Theor. Biol. **31**(2), 331–338 (1971)

A. Hong, G. Chen, J.L. Li, Z.R. Chi, D. Zhang, A flower image retrieval method based on ROI feature. J. Zhejiang Univ. Sci. **5**(7), 764–772 (2004)

H. Hotelling, The generalization of Student's ratio. Ann. Math. Stat. **2**(3), 360–378 (1931)

H. Hotelling, The most predictable criterion. J. Educ. Psychol. **26**(2), 139–142 (1935)

H. Hotelling, Relations between two sets of variables. Biometrika **28**(3/4), 321–377 (1936)

M.K. Hu, Visual pattern recognition by moment invariants. IRE Trans. Inf. Theory **8**(2), 179–187 (1962)

L.-L. Huang, A. Shimizu, H. Kobatake, Robust face detection using Gabor filter features. Pattern Recognit. Lett. **26**, 1641–1649 (2005)

Q. Huang, A.K. Jain, G.C. Stockman, A.J.M. Smucker, Automatic image analysis of plant root structures, in *Proceedings 11th IAPR International Conference on Pattern Recognition, 1992. Conference B: Pattern Recognition Methodology and Systems*, vol. II (1992), pp. 569–572. doi: 10.1109/ICPR.1992.201842

Z.-K. Huang, D.-S. Huang, J.-X. Du, Z.-H. Quan, S.-B. Guo, Bark classification based on Gabor filter features using RBPNN neural network, in *International Conference on Neural Information Processing* (Springer, Heidelberg, 2006), pp. 80–87

P.M. Huff, P. Wilf, E.J. Azumah, Digital future for paleoclimate estimation from fossil leaves? Preliminary results. Palaios **18**(3), 266–274 (2003)

D.L. Hull, The effect of essentialism on taxonomy - two thousand years of stasis (I). Br. J. Philos. Sci. **15**(60), 314–326 (1965a)

D.L. Hull, The effect of essentialism on taxonomy - two thousand years of stasis (II). Br. J. Philos. Sci. **16**, 1–18 (1965b)

J.S. Huxley, *Problems of Relative Growth* (Methuen, London, 1932)

J.S. Huxley, *The New Systematics* (The Systematics Association, London, 1971)

C. Im, H. Nishida, T.L. Kunii, *Recognizing plant species by normalized leaf shapes, in Vision Interface '99* (Trois Rivières, Canada, 1999), pp. 397–404

H. Iwata, S. Niikura, S. Matsuura, Y. Takano, Y. Ukai, Genetic control of root shape at different growth stages in radish (Raphanus sativus L.). Breed. Sci. **54**(2), 117–124 (2004)

R. Jacob, K. Karn, Eye tracking in human-computer interaction and usability research: Ready to deliver the promises, in *The Mind's Eye: Cognitive and Applied Aspects of Eye Movement Research* (2003), pp. 573–603

R.J. Jensen, K.M. Ciofani, L.C. Miramontes, Lines, outlines, and landmarks: Morphometric analyses of leaves of Acer rubrum, Acer saccharinum (Aceraceae) and their hybrid. Taxon **51**(3), 475–492 (2002)

W. Johannsen, *Elemente der exakten Erblichkeitslehre* (G. Fischer, Jena, 1909)

P. Jolicoeur, J.E. Mosimann, Size and shape variation in the painted turtle. a principal component analysis. Growth **24**, 339–354 (1960)

S. Joly, A. Bruneau, Delimiting species boundaries in Rosa Sect. Cinnamomeae (Rosaceae) in eastern North America. Syst. Botany **32**(4), 819–836 (2007)

L. Journaux, M.-F. Destain, J. Miteran, A. Piron, F. Cointault, Texture classification with generalized Fourier descriptors in dimensionality reduction context: An overview exploration. Artif. Neural Netw. Pattern Recognit. **5064**, 280–291 (2008)

F. Jurie, B. Triggs, Creating efficient codebooks for visual recognition, in *Proceedings of the Tenth IEEE International Conference on Computer Vision (ICCV'05)*, vol. 1 (IEEE Computer Society, Washington DC, 2005), pp. 604–610

K. Kavukcuoglu, M. Ranzato, R. Fergus, Y. LeCun, Learning invariant features through topographic filter maps, in *Proceedings of the International Conference on Computer Vision and Pattern Recognition (CVPR'09)* (IEEE, New York, 2009), pp. 1605–1612

D.G. Kendall, The diffusion of shape. Adv. Appl. Probab. **9**(3), 428–430 (1977)

D.G. Kendall, D. Barden, T.K. Carne, H. Le, *Shape and Shape Theory* (Wiley, New York, 1999)

N. Kirchgessner, H. Scharr, U. Schurr, Robust vein extraction on plant leaf images, in *2nd IASTED International Conference Visualisation, Imaging and Image Processing* (2002)

J. Kittler, Feature set search algorithms, in *Pattern Recognition and Signal Processing*, ed. by C.H. Chen (Sijthoff and Noordhoff, Alphen aan den Rijn, Netherlands, 1978), pp. 41–60

C.P. Klingenberg, S. Duttke, S. Whelan, M. Kim, Developmental plasticity, morphological variation and evolvability: a multilevel analysis of morphometric integration in the shape of compound leaves. J. Evolut. Biol. **25**, 115–129 (2012)

S. Knapp, E.N. Lughadha, A. Paton, Taxonomic inflation, species concepts and global species lists. Trends Ecol. Evol. **20**(1) (2005)

D. Koller, M. Sahami, Toward optimal feature selection, in *Proceedings of the 13th International Conference on Machine Learning (ICML)* (1996), pp. 284–292

H.W. Kuhn, The Hungarian method for the assignment problem. Nav. Res. Logist. Q. **2**, 83–97 (1955)

N. Kumar, P.N. Belhumeur, A. Biswas, D. Jacobs, W.J. Kress, I. Lopez, J.V.B. Soares, Leafsnap: a computer vision system for automatic plant species identification, in *Proceedings of the European Conference in Computer Vision (ECCV)* (2012), pp. 497–504

Y. Kurosawa, Incremental learning for feature extraction filter mask used in similar pattern classification, in *International Joint Conference on Neural Networks (IJCNN)* (2008), pp. 497–504

P.E. Lestrel, *Fourier Descriptors and their Applications in Biology* (Cambridge University Press, Cambridge, 1997)

T. Leung, J. Malik, Representing and recognising the visual appearance of materials using three-dimensional textons. Int. J. Comput. Vis. **43**, 7–27 (2001)

H. Lewis, Leaf variation in Delphinium variegatum. Bull. Torrey Bot. Club **74**(1), 57–59 (1947)

C. Lexer, J. Joseph, M. van Loo, G. Prenner, B. Heinze, M.W. Chase, D. Kirkup, The use of digital image-based morphometrics to study the phenotypic mosaic in taxa with porous genomes. Taxon **580**(2), 349–364 (2009)

W. Li, K. Mao, H. Zhang, T. Chai, Designing compact Gabor filter banks for efficient texture feature extraction, in *2010 11th International Conference on Control Automation Robotics Vision (ICARCV)* (2010), pp. 1193–1197

Y. Li, Z. Chi, D.D. Feng, Leaf vein extraction using independent component analysis, in *IEEE International Conference on Systems, Man and Cybernetics* (IEEE, New York, 2006), pp. 3890–3984

M. Lipske, *New Electronic Field Guide Uses Leaf Shapes to Identify Plant Species* (Inside Smithsonian Research, Winter, 2008)

J. Liu, Y.Y. Tang, Adaptive image segmentation with distributed behavior-based agents. IEEE Trans. Pattern Anal. Mach. Intell. **210**(6), 544–551 (1999)

J. Liu, S. Zhang, S. Deng, A method of plant classification based on wavelet transforms and support vector machines. Emerg. Intell. Comput. Technol. Appl. **5754**, 253–260 (2009)

W. Ma, H. Zha, J. Liu, X. Zhang, B. Xiang, Image-based plant modeling by knowing leaves from their apexes, in *19th International Conference on Pattern Recognition, 2008* (2008), pp. 1–4

N. MacLeod, Generalizing and extending the eigenshape method of shape space visualization and analysis, in *Paleobiology* (1999), pp. 107–138

N. MacLeod, *Automated Taxon Identification in Systematics: Theory, Approaches and Applications*. Systematics Association Special vol. 74 (CRC Press, Boca Raton, 2007a)

N. MacLeod, Introduction, in *Automated Taxon Identification in Systematics: Theory, Approaches and Applications*, ed. by N. MacLeod, Systematics Association Special, vol. 74, (CRC Press, Boca Raton, 2007b), pp. 1–7

N. MacLeod, Palaeo-math 101: going round the bend: eigenshape analysis I. Palaeontol. Newsl. **800**(26), 32–48 (2012a)

N. MacLeod, Palaeo-math 101: going round the bend II: extended eigenshape analysis. Palaeontol. Newsl. **810**(27), 23–39 (2012b)

N. MacLeod, Palaeo-math 101: landmarks and semilandmarks: differences without meaning and meaning without difference. Palaeontol. Newsl. **820**(28), 32–43 (2013)

S. Magrini, A. Scoppola, Geometric morphometrics as a tool to resolve taxonomic problems: the case of *Ophioglossum* species (ferns), in *Tools for Identifying Biodiversity: Progress and Problems*, P.L. Nimis, R. Vignes Lebbe, (EUT, The Netherlands, 2010), pp. 251–256

P.C. Mahalanobis, On the generalized distance in statistics. Proc. Natl. Inst. Sci. India **20**(1), 49–55 (1936)

S. Mannan, C. Kennard, M. Husain, The role of visual salience in directing eye movements in visual object agnosia. Current Biol. **19**, 247–248 (2009)

L. Marcus, Traditional morphometrics, in *Proceedings of the Michigan Morphometrics Workshop*, ed. by F.J. Rohlf, F.L. Bookstein (The University of Michigan Museum of Zoology, Ann Arbor, Michigan, 1990), pp. 77–122

K. Marhold, Multivariate morphometrics and its application to monography at specific and infraspecific levels, in *Monographic Plant Systematics, Fundamental Assessment of Plant Biodiversity*, T.F. Stuessy, W. Lack (A.R.G. Gantner Verlag K.G., 2011), pp. 75–101

T. Martinetz, K. Shulten, A Neural-gas network learns topologies, in *Artificial, Neural Networks* (1991), pp. 397–402

K. Mather, J. Jinks, *Biometrical Genetics: The Study of Continuous Variation* (Chapman and Hall, London, 1971)

E. Mayr, Numerical phenetics and taxonomic theory. Syst. Zool. **140**(2), 73–97 (1965)

E. McClintock, C. Epling, A revision of Teucrium in the New World with observations on its variation, geographical distribution, and history. Brittonia **50**(5), 491–510 (1946)

T. McLellan, Geographic variation and plasticity of leaf shape and size in Begonia dregei and B. homonyma (Begoniaceae). Bot. J. Linn. Soc. **1320**(1), 79–95 (2000)

T. McLellan, Correlated evolution of leaf shape and trichomes in Begonia dregei (Begoniaceae). Am. J. Botany **920**(10), 1616–1623 (2005)

T. McLellan, J.A. Endler, The relative success of some methods for measuring and describing the shape of complex objects. Syst. Biol. **47**, 264–281 (1998)

C. Meade, J. Parnell, Multivariate analysis of leaf shape patterns in Asian species of the Uvaria group (Annonaceae). Bot. J. Linn. Soc. **143**, 231–242 (2003)

E.D. Megaw, J. Richardson, Eye movements and industrial inspection. Appl. Ergon. **10**, 145–154 (1979)

R. Melville, The accurate definition of leaf shapes by rectangular coordinates. Ann. Botany, N. Ser. **10**(4), 673–679 (1937)

R. Melville, A joint discussion with the systematics association on biometrics and systematics: On the application of biometrical methods in plant taxonomy. Proc. Linn. Soc. Lon. **1620**(2), 153–159 (1951)

G. Mendel, Versuche über Pflanzen-Hybriden. Verhandlungen des Naturforschenden Vereins in Brunn **4**, 3–47 (1866)

P. Mitteroecker, P. Gunz, Advances in geometric morphometrics. Evolut. Biol. **360**(2), 235–247 (2009)

F. Mokhtarian, Silhoutte based isolated object recognition through curvature scale space. IEEE Trans. Pattern Anal. Mach. Intell. **170**(5), 539–544 (1995)

F. Mokhtarian, S. Abbasi, Matching shapes with self-intersection: application to leaf classification. IEEE Trans. Image Process. **13**, 653–661 (2004)

R. Mullen, D. Monekosso, S. Barman, P. Remagnino, P. Wilkin, Artificial ants to extract leaf outlines and primary venation patterns. Lect. Notes Comput. Sci. **5217**, 251–258 (2008)

Y. Nam, E. Hwang, D. Kim, *CLOVER: a mobile content-based leaf image retrieval system, in Digital Libraries: Implementing Strategies and Sharing Experiences, LNCS*, vol. 3815 (Springer, Heidelberg, 2005), pp. 139–148

M.E. Nilsback, A. Zisserman, Delving into the whorl of flower segmentation. Proc. Br. Mach. Vis. Conf. **1**, 570–579 (2007)

E. Nilsson-Ehle, Kreuzungsuntersuchungen an Hafer und Weizen. *Acta Universitatis Lundensis, Ser.* **250**(2), 1–122 (1909)

D. Oakley, H.J. Falcon-Lang, Morphometric analysis of Cretaceous (Cenomanian) angiosperm woods from the Czech Republic. Rev. Palaeobot. Palynol. **1530**(3–4), 375–385 (2009)

M.A. O'Neill, Daisy: a practical computer-based tool for semi-automated species identification, in *Automated Taxon Identification in Systematics: Theory, Approaches and Applications*, ed. by N. MacLeod (CRC Press, Florida, 2007), pp. 101–114

N. Otsu, A threshold selection method from gray level histograms. IEEE Trans. Syst. Man Cybern. **9**, 62–66 (1979)

J. Pan, Y. He, Recognition of plants by leaves digital image and neural network, in *2008 International Conference on Computer Science and Software Engineering* (Wuhan, China, 2008), pp. 906–910

R.J. Pankhurst, *Practical Taxonomic Computing* (Cambridge University Press, Cambridge, 1991)

J. Park, E. Hwang, Y. Nam, Utilizing venation features for efficient leaf image retrieval. J. Syst. Softw. **810**(1), 71–82 (2008)

E. Parzen, On estimation of a probability density funstion and mode. Ann. Math. Stat. **33**, 1065–1076 (1962)

S.C. Pearce, *Biological Statistics: An Introduction* (McGraw-Hill Book Company, New York, 1965)

K. Pearson, On lines and planes of closest fit to systems of points in space. Philos. Mag. **2**, 559–572 (1901)

K. Pearson, A. Lee, E. Warren, A. Fry, C.D. Fawcett, Mathematical contributions to the theory of evolution IX - On the principle of homotyposis and its relation to heredity, to the variability of the individual, and to that of the race. Part I. Homotyposis in the vegetable kingdom. Philos. Trans. R. Soc. Lond. A **197**, 285–379 (1901)

R.A. Pimentel, *Morphometrics: The Multivariate Analysis of Biological Data* (Kendall/Hunt Publishing Company, Dubuque, 1979)

R.O. Plotze, M. Falvo, J.G. Padua, L.C. Bernacci, M.L.C. Vieira, G.C.X. Oliveira, O. Martinez, Leaf shape analysis using the multiscale Minkowski fractal dimension, a new morphometric method: A study with Passiflora (Passifloraceae). Canadian J. Botany **830**(3), 287–301 (2005)

T.M. Porter, *The Rise of Statistical Thinking 1820–1900* (Princeton University Press, Princeton, 1986)

P. Prusinkiewicz, J.S. Hanan, Lindenmayer systems, fractals, and plants. Lect. Notes Biomath. **79**, (1989)

P. Prusinkiewicz, A. Lindenmayer, *The Algorithmic Beauty of Plants* (Springer, Berlin, 1990)

P. Prusinkiewicz, A. Runions, Computational models of plant development and form. N. Phytol. **1930**(3), 549–569 (2012)

L. Qi, Q. Yang, G. Bao, Y. Xun, L. Zhang, A dynamic threshold segmentation algorithm for cucumber identification in greenhouse, in *2nd International Congress on Image and Signal Processing, 2009* (2009), pp. 1–4

M.H. Quenouille, *Associated Measurements* (Butterworths Scientific Publications, London, 1952)

T. Randen, J.H. Husoy, Filtering for texture classification: a comparative study. IEEE Trans. Pattern Anal. Mach. Intell. **210**(4), 291–310 (1999)

E. Rashedi, H. Nezamabadi, S. Saryazdi, A simultaneous feature adaptation and feature selection method for content-based image retrieval systems. Knowl. Based Syst. **39**, 85–94 (2013)

T.S. Ray, Landmark eigenshape analysis: homologous contours leaf shape in Syngonium (Araceae). Am. J. Botany **79**, 69–76 (1992)

R.L. Mendona, Ernesto Rego, A.F.R. Araujo, F.B. Lima Neto, Growing self-reconstruction maps. IEEE Trans. Neural Netw. **210**(2), 211–223 (2010)

L.W. Renniger, P. Verhese, J. Coughlan, Where to look next? Eye movements reduce local uncertainty. J. Vis. **7**, (2007)

R.A. Reyment, Multivariate morphometrics and analysis of shape. Math. Geol. **170**(6), 591–609 (1985)

R.A. Reyment, *Multidimensional Palaeobiology* (Pergamon Press, Oxford, 1991)

R.A. Reyment, An idiosyncratic history of early morphometrics, in *Advances in Morphometrics*, ed. by L.F. Marcus, M. Corti, A. Loy, G.J.P. Naylor, D.E. Slice (Plenum Press, New York, 1996), pp. 15–22

R.A. Reyment, R.E. Blackith, N.A. Campbell, *Multivariate Morphometrics*, 2nd edn. (Academic Press, London, 1984)

R.A. Richards, *The Species Problem: A Philosophical Analysis* (Cambridge University Press, Cambridge, 2010)

D.C. Richardson, M.J. Spivey, W.G. Daly, *Eye-Tracking: Characteristics and Methods Eye-Tracking: Research Areas and Applications* (Informa Healthcare, Switzerland, 2004)

A. Riedel, K. Sagata, Y.R. Suhardjono, R. Tanzler, M. Balke, Integrative taxonomy on the fast track - towards more sustainability in biodiversity research. Front. Zool. **10**, 1–9 (2013)

O.C. Rieppel, *Fundamentals of Comparative Biology* (Birkhauser Verlag, Basel, 1988)

F.J. Rohlf, Correlated characters in numerical taxonomy. Syst. Zool. **160**(2), 109–126 (1967)

F.J. Rohlf, Relationships among eigenshape analysis, Fourier analysis and analysis of coordinates. Math. Geol. **180**(8), 845–854 (1986)

F.J. Rohlf, *ntsYSpc: Numerical Taxonomy System ver. 2.20d* (Exeter Publishing, Ltd. Setauket, NY., 2005)

F.J. Rohlf, *tpsDig, Digitize Landmarks and Outlines, ver. 2.16* (Stony Brook, NY: Department of Ecology and Evolution, State University of New York, 2010a)

F.J. Rohlf, *tpsRelw, Relative Warps Analysis, ver. 1.49* (Stony Brook, NY: Department of Ecology and Evolution, State University of New York, 2010b)

F.J. Rohlf, *tpsUtil, File Utility Program, ver. 1.46* (Stony Brook, NY: Department of Ecology and Evolution, State University of New York, 2010c)

F.J. Rohlf, *Morphometrics at SUNY Stony Brook* (Stony Brook, NY: Department of Ecology and Evolution, State University of New York, 2014). http://life.bio.sunysb.edu/morph/index.html

F.J. Rohlf, L.F. Marcus, A revolution in morphometrics. Trends Ecol. Evol. **80**(4), 129–132 (1993)

P. Room, J. Hanan, P. Prusinkiewicz, Virtual plants: new perspectives for ecologists, pathologists and agricultural scientists. Trends Plant Sci. **1**, 33–38 (1996)

D.L. Royer, P. Wilf, Why do toothed leaves correlate with cold climates? Gas exchange at leaf margins provides new insights into a classic paleotemperature proxy. Int. J. Plant Sci. **1670**(1), 11–18 (2006)

D.L. Royer, P. Wilf, D.A. Janesko, E.A. Kowalski, D.L. Dilcher, Correlations of climate and plant ecology to leaf size and shape: potential proxies for the fossil record. Am. J. Botany **920**(7), 1141–1151 (2005)

Y. Rubner, L.J. Guibas, C. Tomasi, The earth mover's distance, multi-dimensional scaling, and color-based image retrieval, in *ARPA Image Understanding Workshop* (1997), pp. 661–668

W.J. Ryan, A.T. Duchowski, E.A. Vincent, D. Battisto, Match-moving for area-based analysis of eye movements in natural tasks, in *Proceedings of the 2010 Symposium on Eye-Tracking Research and Applications, ETRA 10* (ACM, New York, 2010), pp. 235–242

H. Sakoe, S. Chiba, Dynamic programming algorithm optimization for spoken word recognition. IEEE Trans. Acoust. Speech Sig. Process. **260**(1), 43–49 (1978)

S. Salvador, P. Chan, Toward accurate dynamic time warping in linear time and space. Intell. Data Anal. **110**(5), 561–580 (2007)

R. Sandler M. Lindenbaum, Gabor filter analysis for texture segmentation, in *Computer Vision and Pattern Recognition Workshop* (IEEE, New York, 2006), p. 178

Y. Savriama, J. Neustupa, C.P. Klingenberg, Geometric morphometrics of symmetry and allometry in Micrasterias rotata, Zygnemophyceae, Viridiplantae). Nova Hedwigia, Beiheft **136**, 43–54 (2010)

D. Šeatović, A segmentation approach in novel real time 3D plant recognition system, in *Proceedings of the 6th International Conference on Computer Vision Systems* (2008), pp. 363–372

S. Sempena, N.U. Maulidevi, P.R. Aryan, Human action recognition using dynamic time warping, in *International Conference on Electrical Engineering and Informatics* (IEEE, New York, 2011), pp. 1–5

A.B. Shipunov, R.M. Bateman, Geometric morphometrics as a tool for understanding *Dactylorhiza* (Orchidaceae) diversity in European Russia. Biol. J. Linn. Society, 850 (1):0 1–12, 2005

R. Sievanen, C. Godin, T.M. DeJong, E. Nikinmaa, Functional-structural plant models: a growing paradigm for plant studies. Ann. Botany **1140**(4), 599–603 (2014)

K. Sikka, T. Wu, J. Susskind, M. Bartlett, Exploring bag of words architectures in the facial expression domain, in *European Conference on Computer Vision* (2012)

M.F.S. Silva, I.M. Andrade, S.J. Mayo, Geometric morphometrics of leaf blade shape in Montrichardia linifera (Araceae) populations from the Rio Parnaiba Delta, north-east Brazil. Bot. J. Linn. Soc. **170**, 554–572 (2012)

J. Sivic, A. Zisserman, Google video: a text retrieval approach to object matching in videos. Int. Conf. Comput. Vis. **2**, 1470–1477 (2003)

D.E. Slice, *Modern Morphometrics in Physical Anthropology* (Springer, New York, 2005)

D.E. Slice, Morpheus et al., Multiplatform software for morphometric research (2008). http://www.morphometrics.org/id6.html

D.E. Slice, F.L. Bookstein, L.F. Marcus, F.J. Rohlf, *A Glossary for Geometric Morphometrics* (Stony Brook, NY: Department of Ecology and Evolution, State University of New York, 2015). http://life.bio.sunysb.edu/morph/. Accessed 9 Jul 2015

P.H.A. Sneath, Recent developments in theoretical and quantitative taxonomy. Syst. Zool. **100**(3), 118–139 (1961)

P.H.A. Sneath, The construction of taxonomic groups, in *Microbial Classification*, ed. by G.C. Ainsworth, P.H.A. Sneath (Cambridge University Press, Cambridge, 1962), pp. 289–332

P.H.A. Sneath, Thirty years of numerical taxonomy. Syst. Biol. **440**(3), 281–298 (1995)

P.H.A. Sneath, R.R. Sokal, *Numerical Taxonomy* (W. H. Freeman and Co., San Francisco, 1973)

P. Soille, Morphological image analysis applied to crop field mapping. Image Vis. Comput. **180**(13), 1025–1032 (2000)

R.R. Sokal, Distance as a measure of taxonomic similarity. Syst. Zool. **100**(2), 70–79 (1961)

R.R. Sokal, Typology and empiricism in taxonomy. J. Theor. Biol. **30**(2), 230–267 (1962)

R.R. Sokal, P.H.A. Sneath, *Principles of Numerical Taxonomy* (W. H. Freeman and Company, San Francisco, 1963)

Yu. Song, R. Wilson, R. Edmondson, N. Parsons, Surface modelling of plants from stereo images, in *Sixth International Conference on 3-D Digital Imaging and Modeling, 2007. 3DIM '07* (2007), pp. 312–319

K. Sparck-Jones, R.M. Needham, Automatic term classifications and retrieval. Inf. Storage Retr. **4**, 91–100 (1968)

P. Sprent, The mathematics of size and shape. Biometrics **280**(1), 23–37 (1972)

W.T. Stearn, *Botanical Latin* (David and Charles, Newton Abbot, 1992)

G.L. Stebbins, *Variation and Evolution in Plants* (Columbia University Press, New York, 1950)

G.L. Stebbins, *Edgar Anderson 1897–1969* (Washington D. C, Biographical Memoir, National Academy of Sciences, 1978)

M.R. Teague, Image analysis via the general theory of moments. J. Opt. Soc. Am. **700**(8), 920–930 (1980)

G. Teissier. Un essai d'analyse factorielle. Les variants sexuels de *Maia squinada*. Biotypologie **7**, 73–96 (1938)

C.H. Teng, Y.T. Kuo, Y.S. Chen, Leaf segmentation, its 3D position estimation and leaf classification from a few images with very close viewpoints, in *International Conference on Image Analysis and Recognition* (2009), pp. 937–946

D.W. Thompson, *On Growth and Form* (Cambridge University Press, Cambridge, 1917)

D.W. Thompson, *On Growth and Form: A*, new edn. (Cambridge University Press, Cambridge, 1942)

P. Tirilly, V. Claveau, P. Gros, Language modeling for bag-of-visual words image categorization, in *International Conference on Content-based Image and Video Retrieval (CIVR)* (ACM, New York, 2008), pp. 249–258

M. Turkan, B. Dulek, I. Onaran, A.E. Cetin, Human face detection in video using edge projections, in *Proceedings of the Society of Photographic Instrumentation Engineers, Visual Information Processing XV*, vol. 6246 (2006), p. 624607

W.B. Turrill, The ecotype concept: a consideration with appreciation and criticism especially of recent trends. N. Phytol. **450**(1), 34–43 (1946)

T. Van der Niet, C.P.E. Zollikofer, M.S.P. Leon, S.D. Johnson, H.P. Linder, Three-dimensional geometric morphometrics for studying floral shape variation. Trends Plant Sci. **150**(8), 423–426 (2010)

M. Varma, A. Zisserman, A statistical approach to material classification using image patch exemplars. IEEE Trans. Pattern Anal. Mach. Intell. **31**, 2032–2047 (2009)

V. Viscosi, A. Cardini, Leaf morphology, taxonomy and geometric morphometrics: a simplified protocol for beginners. PLoS ONE **60**(10), 1–20 (2011)

P.A. Volkova, A.B. Shipunov, Morphological variation of Nymphaea (Nymphaeaceae) in European Russia. Nordic J. Botany **250**(5/6), 329–338 (2007)

X.-F. Wang, D.-S. Huang, J.-X. Du, H. Xu, L. Heutte, Classification of plant leaf images with complicated background. Appl. Math. Comput. **2050**(2), 916–926 (2008)

Z. Wang, Z. Chi, F. Dagan, Q. Wang, Leaf image retrieval with shape features. Adv. Vis. Inf. Syst. **1929**, 41–52 (2000)

Z. Wang, Z. Chi, F. Dagan, Shape based leaf image retrieval. Vis. Image Sig. Process. **150**, 34–43 (2003)

G.W. Weber, F.L. Bookstein, *Virtual Anthropology: A Guide to a New Interdisciplinary Field* (Springer, New York, 2010)

M. Wedel, R. Pieters, A review of eye-tracking research in marketing. Rev. Mark. Res. **4**, 123–147 (2008)

W.F.R. Weldon, The variations occurring in certain Decapod Crustacea. Proc. R. Soc. Lond. **47**, 445–453 (1889)

J.G. West, I.R. Noble, Analyses of digitised leaf images of the Dodonaea viscosa complex in Australia. Taxon **330**(4), 595–613 (1984)

J. White, The plant as a metapopulation. Ann. Rev. Ecol. Syst. **10**, 109–145 (1979)

R. White, H.C. Prentice, T. Verwist, Automated image acquisition and morphometric description. Canadian J. Botany **66**, 450–459 (1988)

S. White, S. Feiner, J. Kopylec, Virtual vouchers: prototyping a mobile augmented reality user interface for botanical species identification, in *IEEE Symposium on 3D User Interfaces* (2006), pp. 119–126

P. Wilkin, A morphometric study of Dioscorea quartiniana (Dioscoreaceae). Kew Bull. **54**, 1–18 (1999)

J.S. Wilkins, *Species: A History of the Idea* (University of California Press, Berkeley, 2009)

J.S. Wilkins, What is a species? Essences and generation. Theory Biosci. **1290**(2/3), 141–148 (2010)

J. Wilson, *Biological Individuality: The Identity and Persistence of Living Entities* (Cambridge University Press, Cambridge, 1999)

J.E. Winston, *Describing Species* (Columbia University Press, New York, 1999)

R.E. Woodson Jr., Some dynamics of leaf variation in Asclepias tuberosa. Ann. Mo. Bot. Gard. **34**, 353–432 (1947)

S.G. Wu, F.S. Bao, E.Y. Xu, Y.-X. Wang, Y.-F. Chang, Q.-L. Xiang, *A leaf recognition algorithm for plant classification using probabilistic neural network, in IEEE International Symposium on Signal Processing and Information Technology* (IEEE, New York, 2007)

L. Ye, E. Keogh, Time series shapelets: A new primitive for data mining, in *IEEE International Conference on Knowledge Discovery and Data Mining* (ACM, New York, 2009), pp. 947–956

Y. Yoshioka, H. Iwata, R. Ohsawa, S. Ninomiya, Analysis of petal shape variation of Primula sieboldii by elliptic fourier descriptors and principal component analysis. Ann. Botany **94**, 1657–1664 (2004)

Y. Yoshioka, K. Ohashi, A. Konuma, H. Iwata, R. Ohsawa, S. Ninomiya, Ability of bumblebees to discriminate differences in the shape of artificial flowers of Primula sieboldii (Primulaceae). Ann. Botany **990**(6), 1175–1182 (2007)

L. Yu, H. Liu, Efficient feature selection via analysis of relevance and redundancy. J. Mach. Learn. Res. **5**, 1205–1224 (2004)

M.L. Zelditch, D. Swiderski, H.D. Sheets, *Geometric Morphometrics for Biologists: A Primer*, 2nd edn. (Elsevier, Amsterdam, 2012)

G. Zeng, S.T. Birchfield, C.E. Wells, Rapid automated detection of roots in minirhizotron images. Mach. Vis. Appl. **210**(3), 309–317 (2008)

Index

© Springer-Verlag GmbH Germany 2017
P. Remagnino et al., *Computational Botany*, DOI 10.1007/978-3-662-53745-9